SCIENCE DEIFIED
& SCIENCE DEFIED

SCIENCE DEIFIED & SCIENCE DEFIED

The Historical Significance of Science in Western Culture

From the Bronze Age to the Beginnings of the Modern Era ca. 3500 B.C. to ca. A.D. 1640

Richard Olson

University of California Press
Berkeley   Los Angeles   London

University of California Press
Berkeley and Los Angeles, California

University of California Press, Ltd.
London, England

Library of Congress Cataloging in Publication Data

Olson, Richard, 1940–
    Science deified & science defied.
      1. Science—History.   2. Science and civilization.
I. Title.   II. Title: Science deified and science defied.
Q125.044   303.4'83   82-40093
                     AACR2

Printed in the United States of America

1 2 3 4 5 6 7 8 9

To Kathy and Karlin

Contents

Acknowledgments

In addition to those many debts acknowledged in the references cited below, I owe special thanks to Tad Beckman of Harvey Mudd College and J. Peter Euben of the University of California at Santa Cruz, two friends with whom I have taught joint courses over the past several years and who have served as patient midwives to virtually all of the ideas presented here. Librarians David Kuhner of the Norman F. Sprague Library of the Claremont Colleges and Elizabeth Wrigley of the Francis Bacon Library in Claremont have made tracking down materials a pleasure; and Douglas Kim, who did preliminary work on the figures, Jean E. Sells who produced the final versions, and Michael Sailor, who did the photographic work, made my life much easier than it might have been.

G. N. Cantor, William Davenport, Roger Hahn, John Heilbron, and Robert Westman all read portions of the manuscript at early stages and offered valuable comments, some of which were heeded; and both Sheila Berg and Lyda Boyer of the University of California Press have been remarkably encouraging and helpful.

Support for this work has been gratefully received from the Faculty Research Committee and Dean B. Samuel Tanenbaum of Harvey Mudd College, through a Willard W. Keith, Jr. Fellowship in the humanities, and from the Arnold L. and Lois S. Graves Foundation.

Richard G. Olson

Harvey Mudd College
Claremont, California
January, 1982

Prologue

It seems tragic that in a period of increasing emphasis on the technical aspects of science, few, if any, of our vehicles of general education seek to explore the roles of scientific knowledge, attitudes, and activities in the development of the Western world. Western culture has long been deeply influenced by scientific products and processes. This has been most obvious in connection with our material culture; but it has been perhaps even more pervasive in the conceptual framework within which religious, political, and aesthetic life has developed. Any attempt to offer a genuine liberal education that ignores this influence is likely to be of very limited worth in today's world.

Through the nineteenth and early twentieth century, it made great sense to study classical antiquity for the insight it could provide into the growth and character of democratic and republican forms of government, and into the natures of imperialism and slavery. Similarly, it made sense to look upon the early modern period in Europe as the birthplace of commercial capitalism and the nation-state. This is so because in the recent past most men and women perceived the most critical problems facing Western societies in terms of confrontations between capitalism and communism, between democratic and totalitarian governments, between nationalism and internationalism, and between various notions of what it means to be free or love liberty.

During the past few decades, however, increasing numbers of people have begun to suggest that a new *science-related* set of central issues is supplementing and even replacing those of the recent past. Not only are critical policy issues more and more often focusing on science and technology—such as weapons systems, energy production, and environmental protection—but the very processes by which our societal decisions are made are changing. Within capitalist and communist economies alike, scientific modes of analysis practiced by technical experts are coming to play increasingly dominant roles in what were once viewed as political or economic decisions. Thus, our traditional categories of governance (democratic and totalitarian), and our traditional economic categories (capitalist and communist), are being modified. Fears about this trend have been most frequently and forcefully voiced by radical visionaries like Herbert Marcuse and Theodore Roszak; but even such an otherwise "respectable" and conservative figure as Dwight D. Eisenhower warned in his "Farewell Address to the American People" in 1961 that we must be alert to the danger "that public policy could itself become the captive of a scientific-technical elite."

It seems worthwhile to look anew at the growth of Western culture with this trend in mind. Classical antiquity then becomes fascinating for the insights it can provide into the emergence of scientific modes of thought, and into the ways that some societies have controlled and limited science and technology. Similarly, early modern Europe becomes of critical importance in connection with a development even more dramatic than that of the nation-state or of capitalism: it is the scene of a radical change in our perceived relationship with nature—one in which humans developed a manipulative mentality and came to see themselves in a new way as masters and controllers of the natural world, rather than as integrated into nature as its victims, stewards, or recipients of its bounty.

This work is an attempt to draw attention to some of the many roles that the scientific tradition has played in the formation of Western culture—which is rapidly engulfing the entire world. It focuses on exploring connections between the traditional foci of liberal studies (religion, politics, philosophy, the arts, and literature) and the scientific tradition.

Though I specifically renounce any claims of "disinterested objectivity," this work is intended neither as a celebration nor as a condemnation of the scientific tradition and its impact upon our lives. My attitude toward the overall influence of scientific thought and activity on our culture is ambivalent—at once admiring and alienated, sorrowful and grateful. I am moved by Aeschylus' Prometheus who admonishes us to "Listen to the griefs of mortal men, their helpless state before I placed intelligence within them, and the use of the mind," and who catalogs the enrichment of human life

brought about by the exercise of scientific rationality and its consequence, technological innovation. But I cannot escape the prophetic
warning from Mary Shelley's *Frankenstein or the Modern Prometheus*, whose monstrous product turns upon its human creator to cry
out, "Remember, I have power. . . . I can make you so wretched that
the light of day will be hateful to you. You are my creator, but I am
your master; . . . obey!"

Relationships Between Science and Other Cultural Traditions: Definitions and a Way of Formulating Issues

1

Since at least as early as 450 B.C., Western intellectuals have periodically shown intense concern over the degree of integration of scientific attitudes, techniques, and ideas into the broader culture from which they emerge. By briefly considering some recent expressions of that concern we can provide a framework for focusing the following historical studies.

THE TWO CULTURES ARGUMENT

C. P. Snow argued in *The Two Cultures* (1959) that at least in modern Britain and the United States, the two elite groups who wield greatest power over public opinion and crucial political and social decisions, the scientists and the literary humanists, are educated in such radically different traditions that they share almost no basic values and goals. Snow's concern about this schism was expressed principally in cold war terms—as a fear that humanistic values might inhibit the full exploitation of scientific and technological knowledge, which alone could assure the victory of the "free" world in competition with the Soviet Union for the friendship of the underdeveloped nations. Whether one sympathizes with Lord Snow's political attitudes and agrees with his specific arguments or not, it is clear that he focused substantial attention on an important general problem. Grant that scientists and literary humanists or technologists, social scientists, and humanists,[1] or any other combination of two, three, or more specialist groups, play major roles in determining important collective social attitudes and actions. Grant also that these special groups are so differently educated and trained that they have fundamentally divergent value assumptions, often use mutually incomprehensible technical languages, and appeal to unshared analogs, metaphors, and symbols to provide a context for the meaning of terms in their nominally shared language. Then there are bound to be tensions within the society based upon conflicts of interest or conflicts of values, and upon misunderstandings or complete incomprehension between specialist groups.

Many of these conflicts will be minor. Humanists and scientists in the United States, for example, bicker over the relative funding of the National Science Foundation and the National Endowment for the Humanities, each group asserting its own self-importance and its right to a larger share of the society's wealth. Some of the conflicts, on the other hand, will be deep and fundamental. One such conflict that has important bearings on a variety of social issues is most easily seen in the confrontation between behavioral psychologists, exemplified by B. F. Skinner, and humanistic psychologists, exemplified by Carl Rogers.[2] At issue between these men (and groups) is the significance of what many have termed the inner life of humanity—our sense of self-awareness, our feelings of freedom to choose between alternative acts, and our capacity for such things as joy, awe, satisfaction, religious faith, and aesthetic appreciation, which many of us feel to be somehow prior to and more fundamental than any observable behavioral patterns associated with these terms. For Skinner and the scientific specialists, behavior is more important than "inner feelings" because it can be predicted and controlled. And because behavior can often be controlled without concern for inner feelings our sense of free choice must be illusory and all inner life may be totally disregarded for practical purposes. For Rogers and the humanists inner life is vastly more important than overt behavior; our sense of freedom to choose is so immediate that no chain of sensory observations and inferences can refute it, and any group that disregards the inner dimension of human existence is misguided. Such differences in attitude have ramifications in virtually all facets of life.

Whether major or minor, each conflict between specialist groups, because it depends upon conflicts of interest or values or upon inability to communicate, will reinforce the tendency of the members of those groups to act toward one another as members of alien cultures. This means they will draw outsider-insider distinctions and will tend to be mistrustful, competitive rather than cooperative, and often even overtly hostile to one another.

According to Snow, the solution to this problem is relatively simple. We need only make sure that many members of our society have at least rudimentary educations in both or all of the critical specialties. If this were to happen, misunderstandings would be minimized as a source of conflict, some crossover of values would probably occur, and members of one specialty would be more sensitive to the interests of others. Conflict would not be eliminated but might be made more manageable and less enervating; and might even be the source of creative innovation upon occasion.

Jacques Monod, winner of the 1965 Nobel Prize in medicine and physiology, saw in the radical disjunction between scientific values and attitudes and those held by the spokesmen for traditional Western values an even greater danger than that envisioned by Snow. According to Monod,

> the traditional concepts which provided the ethical foundation of human societies from time immemorial, all consisted of imaginary ontogenies, none of which remains tenable in the face of scientific enquiry. . . . This contradiction creates intolerable tensions under which [modern] societies will collapse unless a radically new foundation for their value system can be found.[3]

To this radical problem Monod offers a deceptively simple solution. We must provide "A rational reconstruction of our value system" that embraces the findings and attitudes of science.[4] We must thus abandon all sources of cultural tension to ensure social stability; and in that process we must keep science and *eliminate* the mythical and religious sources of competing values.

THE MONOLITHIC CULTURE ARGUMENT

A stark alternative to Snow's analysis of the problem of integrating the scientific specialty into modern society and to Monod's proposal to reshape culture in conformity to science has been presented in a complex tradition of social criticism symbolized by Herbert Marcuse's *One Dimensional Man* (1964) and Jacques Ellul's *The Technological Society* (1964). Though they diverge on many issues, both Ellul and Marcuse begin with fundamentally humanistic biases, which contrast with Snow's and Monod's proscientific biases. Both, moreover, contend that the values, attitudes, products, and methods of science (or in Ellul's terms, "Technique") have already become so thoroughly integrated into modern industrial societies that they threaten to eradicate all competition. The monolithic culture that is emerging from the domination of scientific rationality or technique will certainly lead to social stability, as Monod would wish. This is so both because all grounds from which any criticism of its values and attitudes could be launched are being eroded, and because it is capable of providing a level of material comfort so high that its sharers are unlikely to find any easily expressed wants unsatisfied— they will be aware of no reason to seek change.

The emergent society will not only be static, however, according to Ellul and Marcuse, it is bound to be totalitarian and repugnant in fundamental ways because it represents the disappearance or subordination of humanistic values, which they hold dear. Marcuse focuses on the inevitability of patterns of domination of man by man which he finds particularly abhorrent.[5] Ellul's vision is, if anything, even more depressing than Marcuse's. For while Marcuse sees scientific rationality and its stepchild, technological rationality, as vehicles that support the domination of man by man, Ellul sees man becoming the victim of a runaway scientific-technical mentality that perpetuates itself and dominates men blindly, impersonally, and to no human ends at all.

If either Ellul or Marcuse is correct, it would seem that the last thing we should seek is what Snow explicitly calls for—a further integration and accommodation between scientific and humanistic specialties; or what Monod seeks—the stability of integrated science and ethical values. At one stage Snow's approach might have been able to help; but now science is so strong that humanistic ideas would only be swallowed up by the dominant scientism. Monod's approach is simply unthinkable—a welcoming of the worst possible world. According to Marcuse, humanists offer the only refuge, the only vital space for autonomy and opposition, the only barrier to a scientific and technological totalitarianism. And the only hope for a better world is to preserve and strengthen the opposition specialty until such time that it is able to spearhead a literal revolution against the status quo.[6]

Once more, Ellul seems even less hopeful. "I merely aver," he says, "that in the present social situation there is not even the beginnings of a solution, no breach in the system of technical necessity."[7]

Who is right? Is the conflict between science and humanistic concerns growing so intense that it threatens to become an obstacle to cultural adaptation and the locus for bitter and wasteful hostility?

Are we on the way to social collapse because our ethical systems are incompatible with the findings of scientists? Has modern culture become so scientized that it is on a course toward a kind of stasis characterized by a monotonous but benign totalitarianism of scientific and technological rationality? Or are all of these alternatives so distorted as to represent paranoid responses to an imperfect but basically sound and adequately flexible cultural situation?

EXPLICIT CONFLICT AND IMPLICIT ASSIMILATION

It seems to me that unease, but not panic, about the role of scientific attitudes and activities in our culture is fully warranted and that, paradoxically, both those who fear too little and too much integration of scientific values and attitudes are substantially correct, even though both groups exaggerate their claims. The apparently contradictory claims of Two Culturists and Monolithic Culturists can be reconciled if we recognize a distinction between explicit and implicit cultural values, assumptions, methods, and styles.

Conflicts between scientific and humanistic values and attitudes are easy to see because they are frequently self-conscious and fully articulated. B. F. Skinner and Carl Rogers, for example, make a special point of expressing their conflicting goals and attitudes. Humanists like Ellul, Marcuse, and Theodore Roszak (author of *The Making of the Counter Culture*, 1969, and *Where the Wasteland Ends*, 1976) spend substantial parts of their lives in articulating the differences between their values, assumptions, and methods, and those of modern science and technology. And scientists like Jacob Bronowski (*Science and Values*, 1956) and Isadore Rabi (*Science, the Center of Culture*, 1971) extoll the values and assumptions of science and defend them against what they often consider to be unfair and unwarranted criticism. To most professional intellectuals—to the leaders, or more properly, the public spokesmen for the specialties involved—the conflicts between scientific and humanistic attitudes, activities, and modes of thought are often at least annoying and inescapable facts of life.

When we recognize the awareness of conflict among the intellectual elite in the humanistic and scientific specialties, however, we must also recognize that between 90 and 99 percent of the members of modern industrialized societies fail to fall into these groups. What of the vast majority in our societies? How do their attitudes and actions relate to science?

Just as surely as the life of every lay person in Medieval Europe took much of its tone and direction from a Christian theology whose intellectual intricacies were known to only a few, the life of every citizen in modern society is deeply colored by the processes and products of scientific thinking.[8] Alvin Weinberg was no doubt correct when he wrote, "When history looks at the twentieth century, she will see science and technology as its theme. She will find in the monuments of Big Science—the huge rockets, the high-energy accelerators, the high-flux research reactors—symbols of our time just as surely as Notre Dame is a symbol of the Middle Ages."[9]

The general diffusion of scientific values, attitudes, concepts, vocabulary, and methods to a point of prevalence and near domination in common life, however, is neither as obvious to us as the theological domination of medieval life, nor as immediately visible as the present overt clashes between scientific and humanistic intellectuals. It is all but invisible because it is often indirect, and because it occurs on a level that is usually tacit and unanalyzed.

The vast majority of people in modern industrial societies *do* value and honor science both as a source of material plenty and as a source of intellectual certainty, even though they cannot explain or understand just why it should be so.[10]

Science is thus widely supported in modern society. More importantly, scientists, as technical experts, are entrusted with influencing societal decisions. In this way scientific values come to play an important direct role in our lives. Few of us ever seek to know how because we have simply abdicated any responsibility for those decisions. We look to our experts for advice on new sources of energy, for standards of safety, and for strategies for national defense. In many of these cases a "political" choice is made from options presented by scientific and technical experts. But the very fact that scientists play a dominant role in formulating both problems and policy options means that the terms of discussion are often restricted to fit within the perspective of the scientific specializations in our society.

Still more important and pervasive are the ways that scientific attitudes and theories structure our ways of thinking about a whole range of issues that are widely held to be quite independent of science. Almost all contemporary Americans, for example, think about our government as something instituted primarily to protect the private interests of individual citizens. Yet only a handful are aware that such a notion of government would have been unthinkable until it was formulated during the seventeenth century in connection with the growth of mechanistic natural philosophy. Even the democratic egalitarianism of our political ideology, with its ideal of equality of opportunity for all, though it has important early Christian roots, was first fully formulated and articulated in connection with the now outdated principles of eighteenth-century sensationalist psychology.

These comments are *not* made to denigrate or undermine our beliefs in individualism and equality of opportunity, nor do they suggest that in all, or most, cases the advice of scientific and technical experts should be ignored in the course of making social decisions. Instead, they *are* made to suggest that most of us are ignorant of the degree to which modern culture has assimilated and been moulded by scientific attitudes and ideas. To the extent that we remain ignorant and blindly accept what seems to be a dominant cultural trend, we do what Ellul fears—we abdicate responsibility for our values and we ensure the continuance of that trend. However, if we seek to understand how scientific notions have come to inform Western culture and what alternatives might exist, we at least make possible some conscious choice on an individual level, which might eventually form a basis for intentional collective action.

A DEFINITION OF SCIENCE

Up to this point I have assumed that everyone has an adequate idea of what "science" is and what special values, attitudes, and methods it presupposes and promotes. In addition, I have implied that science is somehow opposed to "the humanities" and I have suggested that Jacques Ellul's notion of "Technique" is identical with or akin to my own notion of "science." While each of these assumptions, implications, and suggestions is appropriate to the very broad ends I have had so far, they would lead to confusions, misunderstandings and distortions in the more detailed discussions that follow.

For the following discussions, science is taken to be *a set of activities and habits of mind aimed at contributing to an organized, universally valid, and testable body of knowledge about phenomena.* At any given time, these general characteristics are usually embodied in systems of concepts, rules of procedure, theories, and/or model

investigations that are generally accepted by groups of practitioners—the scientific specialists.

Activities we now naturally see as aspects of a scientific specialty were almost certainly seen in many earlier cultures as integral
components of craft traditions (we would probably want to call these
technological specialties), religious traditions, magical traditions,
or broadly philosophical traditions. In spite of this fact, we can see
that activities and ideas that closely resemble modern science in
many of their salient features have existed and been important almost since the beginnings of a written historical record. Those concerned with the heavenly bodies in ancient Mesopotamia combined
religious, magical, and agricultural functions; but their observational and computational activities and their predictive aims had
much in common with those of modern positional astronomers, for
example. Similarly, Ionian Greek sages like Thales of Miletus and
Democritus of Abdera were simultaneously governmental officials,
military officers, commercial entrepreneurs, and moral and natural
philosophers; but their attempts to understand the order and com-
position of the cosmos share many of the fundamental aims of
modern cosmologists and elementary particle physicists. Moreover,
the scientific aspects of the thought of these men have had a profound direct impact upon Western intellectual life, as well as a less
direct but equally important impact upon our social institutions.

When we consider the speculations of Ionian thinkers as scientific, however, we must confront the need to use the term in a very
broad sense. For while these men often sought to establish organized
and universally valid knowledge of phenomena, and while Thales,
for example, even focused on the mathematical relationships between phenomena (this is a pervasive characteristic of modern science) their concepts, theories, and rules of procedure often diverged
substantially from those accepted by modern scientists.

SCIENCE AS A SEARCH FOR ORDER

This brings us back to the problem of indicating more precisely and
concretely the kinds of shared habits of mind and activities that
constitute science. First, science in all of its various branches is
concerned with ordered and systematized knowledge. According to a
dominant modern view, the search for order is biologically adaptive.
It allows us to select from myriads of experiences those we can

comprehend and control; thus people with dominantly order-seeking minds, and cultural specialties with dominantly order-seeking functions, have been favored in the process of natural selection. An ascerbic variant of this view was presented by Henri Poincaré in a statement that helps to show what implications it has for scientific attitudes.

> Instinct is routine, and if thought did not fecundate it, it would no more progress in man than in the bee or ant. It is needful then to think for those who love not thinking and, as they are numerous, it is needful that each of our thoughts be as often useful as possible, and this is why a law will be the more precious the more general it is.
> This shows us how we should choose: the most interesting facts are those which may serve many times; these are the facts which have a chance of coming up again.[11]

According to this view, science focuses upon the repetitive, regular, uniform, and predictable aspects of experience because these aspects allow for *foresight* and planning—for establishing learned and intentional reactions to situations to supplement and supplant instinctual reactions. It seeks increasingly abstract and generalized knowledge because unless we can find ways to consolidate our knowledge of events into small numbers of "laws," or general statements, we must inevitably become buried in countless bits of unconnected information that are useless because they cannot be rapidly recovered and applied. In some sense, then, science seeks systematic and organized knowledge to make our thought economical and efficient for utilitarian purposes.

It is because science often seeks a kind of knowledge that can be exploited for practical ends, and because it highly values the efficient organization of knowledge, just as technology highly values the efficient utilization of resources (human as well as material), that it is easy and frequently appropriate to identify scientific aims and values with those of technology.

SCIENCE AS A SEARCH FOR UNDERLYING FORM

In addition to practical justifications, there are other reasons for the emphasis that scientists place on order and system—reasons that have more to do with psychological and aesthetic considerations than with utility. And these reasons, too, shape the habits of mind associated with science. Again, Henri Poincaré offers important insights into the scientist's drive for order.

> If nature were not beautiful, it would not be worth knowing, and if nature were not worth knowing, life would not be worth living. Of course I do not here speak of that beauty which strikes the senses, the beauty of qualities and of appearances; not that I undervalue such beauty, far from it, but *it has nothing to do with science*; I mean that profounder beauty which comes from the harmonious order of the parts and which a pure intelligence can grasp. This it is which gives body, a structure so to speak, to the irridescent appearances which flatter our senses, and without this support the beauty of these fugitive dreams would only be imperfect, because it would be vague and always fleeting. On the contrary, *intellectual beauty is sufficient unto itself*, and it is for its sake, more perhaps than for the future good of humanity, that the scientist devotes himself to long and difficult labours.[12]

Even though few scientists express the almost mystical rapture of Poincaré, many seek (and virtually all seem to acknowledge and appreciate) a kind of beauty that lies beyond or beneath immediate sensory experience. Furthermore, they attribute to this "intellectual beauty" a kind of self-justification that is denied to immediate emotional responses to qualities and appearances. This special appreciation for the underlying forms of experience is almost always couched in terms of harmony, precision, simplicity and elegance— terms that expose the classical mathematical source of the aesthetic impulse that it defines and responds to.

The fact that scientists usually value an *intellectual* understanding of underlying form over an immediate emotional response to directly experienced objects and events, and the fact that the scientific aesthetic tends to drive scientists to mathematical formulations that can express the kinds of harmony they seek, give to science some of its most important and distinctive characteristics. There have, of course, been scientists who denied the fundamental status of mathematics within science. Aristotle, for example, relegated mathematics to the role of a minor science. Yet he acknowledged that science should seek the underlying form of experience. We regard "artists as *wiser* than persons of mere experience," he wrote, "because the former class know the cause of the thing, the latter not. Persons of mere experience know the *that* but not the why: the others recognize the *why* and the cause."[13] Thus, even in those few sciences whose organizational principles are taxonomic and classificatory rather than mathematical, there is an overriding search for intellectual coherence, pattern, and harmony.

The highly intellectualized and quantified approach to the world that characterizes science has long provided the focus for its harshest critics. The recurrent tensions and conflicts between science and a "romantic" aesthetic (which values emotional and immediate responses to concrete experience rather than intellectual attempts to discover underlying structures and harmonies, and values uniqueness, spontaneity and diversity rather than the universal, the regular and the uniform) will form a continuing theme of this book.

SCIENCE AS A SEARCH FOR UNIVERSALLY VALID AND TESTABLE KNOWLEDGE

Closely associated with the scientific search for order and underlying form is its insistence upon universally valid knowledge—knowledge that is at once certain and independent of the context in which it is discovered or used. One modern tradition of thought, derived from German idealism and associated with the dialectical materialism of Marx and Engels, claims that the underlying forms of experience, or what some call scientific or natural laws, constantly undergo temporal change and development, making them transient rather than universal. Yet dialectical materialists continue to claim that they are scientific and can attain "objective" knowledge.[14] Another very important tradition of thought adopts a radical, skeptical position and claims that scientific knowledge must be couched in provisional theories whose correspondence with phenomena is inevitably limited in scope. In spite of such formal admissions, the search for universal validity has played an important role in the past and continues—as a kind of admittedly unattainable ideal like the Christian admonition to love thy neighbor as thyself—to play an important role in forming the activities and habits of mind of most Western scientists.

The ideal of universal validity implies that scientific knowledge must be available, and the same, to all persons with sufficient intellectual training to understand its meaning, and it must be recognizably valid and accepted as such by those competent to judge. Furthermore, science does not seek to evoke highly personal, emotional, and aesthetic responses; it seeks to be completely independent of them because they are private and highly dependent upon contextual and individual circumstances. In common language, they are highly subjective rather than objective because they depend on the condition of the experienc*er* as well as on that that is experienced. It is this claim to be independent of the spiritual state of the practitioner that distinguishes the science of chemistry from the major tradition of alchemical studies in the Renaissance, for example.

The problem of how one recognizes the validity of purportedly scientific knowledge is an extremely complex one about which philosophers of science are at odds even in the twentieth century, and about which a variety of historically specific positions have been held. At all times and places two general requirements have been among the *minimal* criteria to be met by scientific knowledge:

1. Any system of scientific knowledge must be internally coherent (i.e., no system of scientific statements may contain mutually contradictory statements).
2. Any system of scientific knowledge must be consistent with sensory experience, properly interpreted.

Thus, it has always been agreed that scientific knowledge can be tested against logical criteria and against experiential criteria.

From time to time these different kinds of criteria have been weighted quite unequally. Plato and Platonists, for example, were vastly more concerned with logical coherence than with precise experiential correspondence, whereas Baconian scientists in the seventeenth century were vastly more concerned with experimental tests of scientific statements. Furthermore, groups of scientific specialists have often added more stringent criteria to those minimal ones listed above. Aristotle and his followers, for example, demanded that scientific statements be philosophically necessary (i.e., that they be logically derivable from unimpeachable first principles). A substantial number of seventeenth-century scientists demanded that the causes of all natural effects be conceived in terms of "mechanical motions."[15] And in the late nineteenth and early twentieth centuries many scientists followed Ernst Mach and Percy Bridgeman in demanding that no concept that could not be given an immediate "operational definition" be allowed a role in scientific theorizing.

None of these *special* demands can be allowed to restrict our notion of science, for they would make its scope too historically limited and uninteresting; but we must keep in mind the very general demands for logical coherence and experiential correspondence.

THE SCIENTIFIC EMPHASIS ON PHENOMENA AND QUANTIFICATION

For the time being we assume that the scientific emphasis on sensory experience is not problematic; but the emphasis on logical coherence calls for a brief bit of reflection. For a variety of reasons to be considered later, the domain of mathematical knowledge was among the earliest to be cast into a rigorous deductive and logically coherent form. The propositions of Euclid's *Elements*, for example, followed from the definitions, postulates, and axioms with an unchallengeable logical rigor and certitude in such a way that they had to be accepted as soon as they were understood. For this reason among others mathematical knowledge provided an important model for subsequent attempts at establishing scientific knowledge. And it has been widely—though not universally—agreed that if one could cast the knowledge of any subject matter into mathematical form, then that knowledge would have the logical coherence demanded of all science.

An extreme but illuminating statement by David Hume will allow us to see how the scientific emphasis on mathematical reasoning and sensory experience as sources of knowledge has an important impact upon the range of subjects that science accepts into and excludes from its domain. At the end of his *Enquiry Concerning Human Understanding*, Hume summed up his conclusions by urging men to cull through their libraries, asking of each book on religion, for example, "Does it contain any abstract reasoning concerning quantity or number? No. Does it contain any experimental reasoning concerning matters of fact or existence? No. Commit it then to the flames: for it can contain nothing but sophistry and illusion."[16]

In this statement Hume shows why scientific modes of understanding are conventionally excluded from many topics of religious concern and from metaphysics, which is defined in the dominant Western philosophical tradition as the search for knowledge of a noumenal *reality* assumed to exist beneath the sensory world of phenomena. But in addition Hume demonstrates a kind of intolerance and narrowness of vision more virulent than that of most scientists but similar in its character—a narrowness of vision that threatens to infect us all because it has made possible some remarkable successes within scientific specialties.

Even if we could admit that mathematical reasoning and experimental considerations form the only basis for valid and universal knowledge, and hence for science, we might not agree that the domains of established validity and those of human importance and significance completely overlap. Our gropings with qualitative and affective aspects of life may be incapable of achieving the certitude or coherence of scientific knowledge of phenomena, but they may still be more than sophistry and illusions. Even if they are illusory they may deserve better than to be ignored or discarded because they may be necessary to sustain a meaningful or satisfying human life.

No habit of mind is more characteristically scientific than one that seeks to separate "objective" knowledge from the taint of subjective biases. Such a habit of mind clearly distinguishes scientific from humanistic attitudes, and receives careful scrutiny in the historical chapters of this book.

Although our definition of science started out with an assertion that science is a set of activities and habits of mind, we have focused in this chapter almost exclusively on habits of mind. This is because on the level of mental habits science shows the greatest continuity over time and across subspecialties, and is most easily distinguished from other endeavors; and on this level science has had its most important impact upon culture, and has become implicitly dominant and the source of explicit opposition. But no single habit of mind is exclusively scientific. The search for order in a chaotic world, for example, informs the poetic work as surely as it does the work of science, visual artists struggle for harmony of form, and the desire for universality and certainty is as central to metaphysical and theological speculation as it is to scientific thought. Even if we find that the entire array of desires for system and order, for underlying form, for intellectuality, for quantification, for universality, and for an emphasis on phenomena to the exclusion of intuitive and affective aspects of experience pervade our society, there is no a priori certainty that they do so *because* they are or have been associated with a scientific tradition. Unless we can show that elements of the scientific mentality have entered (or at least become prominent in) our culture in connection with the processes and products of unquestionably scientific *activities* it would be wrong to claim that this book deals with science and culture at all.

Scientific activities have varied so greatly from time to time, place to place, and subdiscipline to subdiscipline that their description and analysis are reserved for the historical part of this work. There, a fundamental task is to demonstrate both that—and how— the habits of mind that we have associated with science have been brought to the attention of broad segments of society and turned into powerful cultural forces in connection with specific scientific activities through the products and popular accounts of scientific methods, theories, and "discoveries."

The Origins and Impact of Scientific Specializations in Ancient Near Eastern Culture

2

THE RISE OF CITIES AND THE THRUST TOWARD CULTURAL SPECIALIZATION

Prior to the rise of urban civilizations based on irrigation agriculture during the fifth and fourth millennia B.C., specialization within human societies was extremely limited. There were undoubtedly age- and sex-based differentiations of roles among both Paleolithic hunter-gatherers and early Neolithic villagers, who were primarily dry farmers and herders. But most members of these earliest societies took part in the full range of cultural activities appropriate to their age and sex. Most adult males in a Paleolithic community were hunters and toolmakers; most women were food gatherers, cooks, and clothes makers; and every head of a family was a religious leader.

A greater degree of specialization began to appear in a few exceptionally large villages and towns of the early Neolithic cultures—at places like Jericho from about 8000 B.C. and at Catal Hüyük in Anatolia from about 6500 B.C. to 5650 B.C. At such centers for the first time there were substantial religious temples whose nature implies the existence of an emerging priesthood; and there were separate workshops for butchers, breadmakers, beadmakers, and bone toolmakers, which indicate substantial economic specialization.[1] It was only with the rise of huge urban centers in southern Mesopotamia around 4000 B.C., however, that cultural specialization and the attendant differentiation of social classes became almost the defining characteristics of Western civilization.

Since geographical and economic factors led directly to the urban life and specialization that ultimately produced science in

16

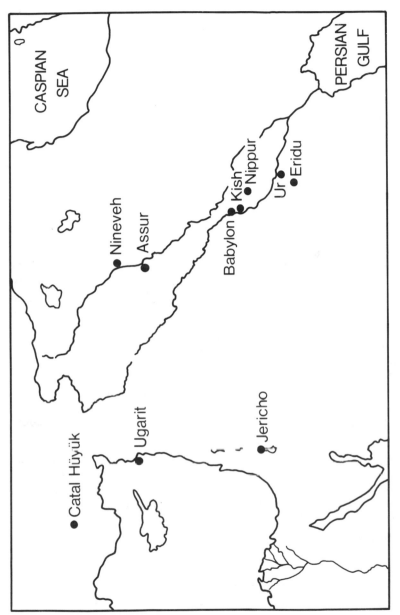

FIG. 1.　Map of the Ancient Near East.

Mesopotamia we must very briefly consider those factors. Small farming villages had existed along the banks of the Euphrates river (see fig. 1) from about 6500 B.C. in the north, and from about 5500 B.C. in the south where the river poured into the Red Sea. The early southern villages were inhabited by a farming and sheep-tending people we have come to call Sumerians. These people characteristically constructed their buildings on reed mats laid on the ground near the marshes and lagoons formed by the silt deposited near the mouth of the river; and they organized their villages around large temple structures.[2] In time, the heavy deposits of silt carried by the Euphrates filled in the marshes, leaving the regions of early settlement on dry land. At the same time climatic changes gradually occurred turning the formerly wet regions near the river into desert lands.

One might expect that a people used to building in marshy areas would simply have abandoned their villages to follow the river mouth or the rainfall as it moved northward; but this did not happen for a simple technical reason. By as early as 5000 B.C. the Sumerians had begun to manipulate the water in their fields by draining marshy plots. And because in many places the natural siltation process built up the bed and banks of the river above the level of the surrounding land, it was easy for the Sumerians to begin to flood or irrigate their dry fields as well as to drain the wet ones.

The advent of irrigation not only made it unnecessary to follow the marshes or the rain, it also made possible the planting and harvesting of two crops each year, almost doubling the agricultural yield. Because the soil was rich large food surpluses could be produced, and this led rapidly to a number of dramatic consequences. First, it led to great population increases. Second, it made possible the freeing of a substantial fraction of the population from involvement in primary food production and allowed them to specialize in providing a variety of other goods and services. Finally, it produced enough surplus wealth to support large-scale trade with outsiders for materials—both necessities and luxuries—that were not locally available.

In recent years there has been an intense (and as yet inconclusive) debate over whether the huge size and complex social organization of the first great cities of Mesopotamia resulted as a response to the demands of large-scale irrigation agriculture, or whether the potential for a highly developed and tightly organized agricultural economy followed from the social organization of substantial cities, which grew up first as ceremonial centers.[3] Regardless of which came first (concentrations of social power or technological change), by the end of the third millennium, several early Sumerian villages had grown into huge urban centers whose populations may have reached between 50,000 and 250,000 persons. Within these cities the temple and priestly hierarchies, which had already begun to emerge in the village cultures, had expanded their activities to organize such things as food and goods distribution systems, police and judicial organizations, and sanitation systems. Above all they had appropriated the power to plan, build, maintain, and supervise the use of the complex and massive hydraulic systems necessary to support the populations—systems in which individual central canals reached seventy-five feet in width and several miles in length, and in which hundreds, even thousands, of mutually interdependent smaller channels existed. The great cities of Mesopotamia thus began as theocracies in which great economic and political as well as sacred authority inhered in the priesthood.

In theory all land surrounding a Sumerian city belonged to the city's god or gods. It was held in common and administered by the priesthood through the head of the temple or the "chief tenant farmer of the god." On the god's behalf small plots were allocated for the use of almost all individuals; thus, most citizens including the priests themselves were at least part-time farmers. But much of the land was reserved to be worked in common for the god[4] by professional full-time farmers. From the temple's wealth, generated both from tithes paid by individual farmers and herdsmen for the use of their plots and from the produce of lands worked directly for the temple, came the produce to support the vast array of specialists who were freed from full-time farming to serve the god and the city in other ways.

Weaving, potting, baking, butchering, leather working, implement making, carpentry, brickmaking, and numerous other crafts, which had once been the common domain of all members of society, gradually became the work of guilds of experts supported by the temples. A whole class of important professionals and intellectuals, the prototypes of merchants, bankers, tax collectors, lawyers, judges, physicians, teachers, clergy, military officers, and so on, came into existence in order to see that goods were collected from those who produced them and distributed to those who needed them, to keep track of debts and services owed to the temple, to adjudicate disputes between individuals or groups, to teach future generations of priests and professionals, to minister to the spiritual needs and physical ailments of the populace, to punish transgressions against the city and its gods, and to lead citizen armies in defense of the city and its lands or in wars of conquest.

Over and above providing circumstances that promoted the growth and differentiation of most of the trades, crafts, and professions central to complex urban cultures even today, the wealth of the Sumerian city temples was sufficient to supply a surplus to trade for materials not available locally. Sumer was rich in lands for tillage and pasturage, but was poor in flint for the production of implements, in semiprecious stones for the beautification of persons and adornment of the temples, and in metals, which were originally used for ornamentation and gradually became fundamental for the production of tools and armaments. Thus, not only were the Sumerians able to purchase and utilize virtually all materials known anywhere, their cities became centers of trade with regions as far distant as East Africa, India, and western Europe. And elements of Sumerian culture were carried with trade goods throughout the Mediterranean basin and beyond into the valleys of the Indus, the Nile, and even the Danube.

Finally, the wealth provided by Mesopotamian irrigation agriculture was used to produce the monumental temple architecture that was one of the most conspicuous characteristics of Sumerian culture, and to produce devotional artifacts of semiprecious stones, ivory, and precious metals, whose beauty of design and fineness of craftsmanship have seldom been matched or surpassed.

THE CONTINUITY OF URBAN CULTURE IN ANCIENT MESOPOTAMIA

By the beginning of the third millennium B.C. most of the basic institutions and patterns of institutional relationships that characterized Mesopotamian culture into the Christian era had been established. There were of course some significant modifications. During the mid-third millennium the functions of the head priest became more and more associated with secular rule; or perhaps more accurately, strong secular leaders usurped some of the authority claimed by the priesthood. In any event, the old appellation of "chief tenant farmer of the god" was supplanted by "lul-gan," "great-man," or king; while the king remained the servant of the city's god, his sacred functions were reduced and he became concerned primarily with judicial and military affairs. By around 2400 B.C., numerous cities were unified periodically under the power of particularly strong kings. At the same time, individual land ownership increasingly tended to replace communal ownership through the temples as the kings appropriated land for themselves and their military supporters. Individual entrepreneurs also began to function more frequently as merchants and traders, supplementing the agents of the great temples; and members of craft guilds became dissociated from specific temples and began to sell their services to secular leaders, temples, and wealthy individuals alike. Surprisingly, even the guild of priests became professionalized, and priests moved from city to city—even perhaps from the service of one god to another.

These changes had some major implications for the culture as a whole. On top of the old primary allegiances to clan, temple, and city, which encouraged men and women to think of themselves first as people of a certain god and of a certain city, there was a growing overlay of new considerations that made people think of themselves fundamentally as scribes, farmers, herders, physicians, merchants, masons, metal workers, and so on. Social class distinctions based principally upon occupational roles began to emerge. The specialization of economic function coupled with the growing secularization of political and economic activities thus tended to reduce the elements of experience and outlook shared by all members of the culture and to enhance the importance of specialized work-related experiences and attitudes in determining even extraoccupational aspects of life.

In spite of these changes, the basic world view of nearly all Sumerians remained tied to traditional beliefs about the major gods; most economic activity continued to be dominated by the temples or the similarly bureaucratized secular courts, and the activities of secular leaders were largely directed (or at least bounded) by religious beliefs and priestly advice. Most importantly for our purposes, nearly all intellectual life, including the education of professionals of all kinds, was tied to institutions whose basic features and outlook had been formulated during the late fourth and early third millennia. These institutions changed remarkably little over 4000 years.

Since it was among the literate intellectual classes, especially among the various kinds of scribes, and within the temples that Mesopotamian scientific activities and ideas came into existence and spread their influence from their original economic context into that of religion and social activity, we soon focus on Sumerian intellectual life in more detail, but one important set of comments is in order first.

In political terms, Sumerian culture ceased to exist during the late third millennium when Akkadian invaders from the north conquered the great cities. From 2000 B.C. the political center of Mesopotamia was usually in Babylon, which was to the north of the great early cities of Sumer. Later still, other invading groups, including the Hittites and Persians, came to rule in parts of Mesopotamia; and with each invasion new languages were introduced to the region and some aspects of the old culture were abandoned. Each of the invading groups, however, tended to incorporate many of the religious and educational elements of the Sumerian cultures. Even when temples were destroyed by invading armies, they were usually rebuilt and rededicated to the old gods; and the worship of Sumerian gods (or of the newcomers' local gods with Sumerian rituals and an accretion of Sumerian associations) supplemented or supplanted the traditional worship of the invaders. Until almost Christian times Sumerian remained the basic language for sacred ritual throughout Mesopotamia, in spite of the fact that from about 2000 B.C. it was a dead language in terms of secular usage.

As the invaders brought new languages to the region they were written in the cuneiform (wedge-shaped) characters of Sumerian script; the Sumerian scribes translated their learning into the new languages, and the learning rapidly became the core of the intellectual heritage of the invaders. The continuity of the educational institutions or schools in which this work was carried on is truly amazing. Excavations at Ur show that as late as about 550 B.C. there was a school or "tablet-house" on precisely the same site that its first ancestor had occupied a thousand years earlier,[5] and the content of the curriculum in both 1500 B.C. and 550 B.C. was basically unchanged (except for the technical content of mathematics exercises) from that followed by aspiring scribes at Shuruppak, where "textbooks" from the mid-third millennium have been found.[6] Thus, even though dramatic political and economic changes occurred through the nearly four millennia that separate ancient Sumer from Babylon at the beginning of the Christian era, the roots, context, and fundamental content of Sumerian intellectual life made important contributions to the intellectual institutions that persisted through Alexandrian times, and even into the period of the Roman empire in the Near East. These intellectual institutions are uniquely important for our purposes, for within them lie the roots of the Western scientific tradition as well as the first examples of the incursion of scientific ideas into important domains of social life.

SUMERIAN LANGUAGE, WRITING, AND EDUCATION

Sometime around 3500 B.C. Sumerian temple functionaries began to develop what was undoubtedly the most important human invention since the origin of language and the control of fire. They started to produce writing in order to keep track of temple inventories and transactions. The first written documents are simply drawings, or pictograms, incised upon small clay tablets to represent commodities and followed by marks to indicate a number of standard measures. At this stage the contents of the tablets can hardly be called writing at all; to be able to draw pictures of things and to accompany them by tally marks are hardly unique accomplishments among humans. No language is composed merely of nouns and numbers; to have true writing one must be able to record such things as actions and states of being, which are either impossible or extremely difficult to convey rapidly and unambiguously with pictorial representations. The Sumerians were the first to develop a real written language out of their pictorial records. They were probably able to do so because of certain special characteristics of their spoken language, and they were almost certainly encouraged in their task by the specific medium in which they kept their records.

Sumerian, like Chinese, Turkish, Hungarian, and Finnish, but unlike most other Near Eastern languages, is an agglutinative language. That is, most simple ideas and everyday objects are expressed in monosyllables; and most complicated or novel meanings are expressed by stringing together an appropriate collection of simple words with no change or inflection. Thus, for example, the Sumerian word for king is big-man or great-man and the word for lapis-lazuli, an important semiprecious stone, is composed of the words for blue, stone, and hard strung together. In addition, Sumerian is full of what we today call homonyms (words that are sounded identically but have vastly different meanings). For example, the sound *ti* meant both "arrow" and "life." In the earliest Sumerian tablets—before about 3200 B.C.—it is fairly clear that the pictograms stood solely for objects; but because those objects could be sounded, and because the sounding could be interpreted as the homonym, it was possible

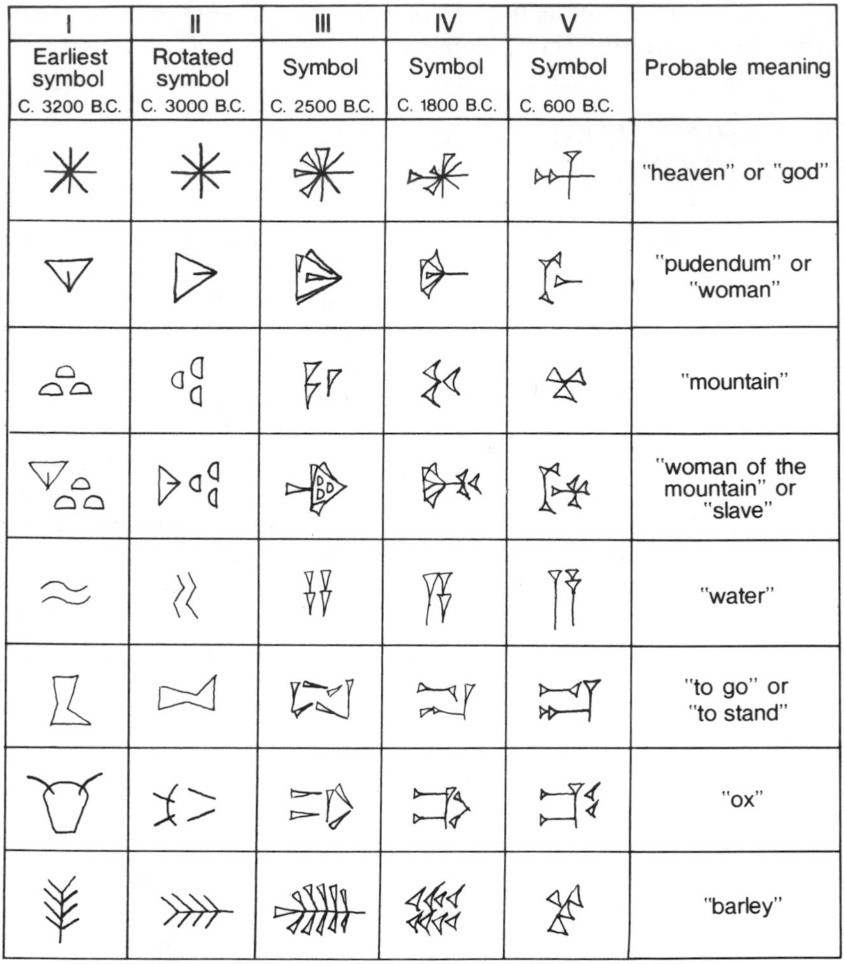

I	II	III	IV	V	
Earliest symbol C. 3200 B.C.	Rotated symbol C. 3000 B.C.	Symbol C. 2500 B.C.	Symbol C. 1800 B.C.	Symbol C. 600 B.C.	Probable meaning
					"heaven" or "god"
					"pudendum" or "woman"
					"mountain"
					"woman of the mountain" or "slave"
					"water"
					"to go" or "to stand"
					"ox"
					"barley"

FIG. 2. Development of cuneiform writing.

to play a kind of game in which the picture stood for the homonym—
or rather, in which the picture stood for a sound whose meaning was
determined by the context. Thus, for example, the pictogram for
sheep accompanied by the "arrow" pictogram could potentially be
read as "live sheep," "birth of a sheep," or "to live like a sheep," as
well as "to kill a sheep with an arrow," and its interpretation would
be apparent from the context and from the presence of other "sounds"

which would further modify the root word for sheep. In this way the original pictograms came to stand for sounds rather than for objects, and a variety of linguistic elements other than nouns were formed by stringing together syllables or sounds. This reinterpretation of pictograms was probably helped along by the fact that making pictograms by simply pressing the end or edge of a stick in the clay writing surface several times was much simpler, faster, and easier than scratching continuous lines to draw a representation of the object. Sumerian pictograms thus rapidly became conventional sets of cuneiform, or wedge-shaped, marks whose likeness to the original object represented was very small. So there was little in the character of the cuneiform marks themselves to inhibit the reading of the pictogram for *ti* as "life" rather than "arrow."

The tremendous practical success of cuneiform writing led to its immediate and widespread use for record keeping. Since the last two decades of the nineteenth century tens of thousands of such things as inventory lists, promissory notes and receipts, deeds of sale, marriage contracts, wills and court decisions have been unearthed from the ruins of ancient Sumerian cities. The importance of record keeping meant that persons trained to read and write were needed; educational institutions were created, and tens of thousands of textbooks and school exercise tablets, a few dating from as early as 3000 B.C. and many from about 2000 B.C. on, provide a detailed record of what Sumerian and later Mesopotamian education was like.

The character of Mesopotamian education and intellectual life in general was to a great extent determined by two factors: first, by the nature of the language itself, and second, by the practical and professional motives for learning. The widespread use of written documents within Mesopotamian culture meant that there was a large demand for literate men, or scribes, in the temples, royal courts, commercial institutions, and law courts. There were, then, many scribes. We know the names of almost a thousand of them from the documents (only one among the thousand recorded was a woman), and they must have formed a substantial part of the population (perhaps several percent of the total). At the same time, since early Sumerian was a syllabic rather than alphabetic language and used up to nearly a thousand different signs or pictograms, education was a long and expensive process; thus, records also show that the vast majority of scribes were sons of professionals and wealthy families. Educated men formed a relatively large and relatively wealthy class in Sumer, and because of the importance of their work they were an influential class.

Because learning was principally a prerequisite for professional careers, there was very little overt encouragement of knowledge for its own sake. The educational curriculum concentrated on basic reading, writing and computational skills and the working of practically oriented problems and exercises. This emphasis is apparent from the general character of most of the exercise tablets that have been excavated, but it is particularly well illustrated in a pedagogical essay that was widely distributed and copied throughout Sumer sometime around 2000 B.C. In this essay two advanced students are fighting over which is the better scholar; in the process they provide a catalog of the desired attainments of a scribe.

The essay begins as one student speaks to the other.

> . . .you Sumerian ignoramus, your hand is terrible; it cannot even hold the stylus properly; it is unfit for writing and cannot take dictation. (And yet you say) you are a scribe like me.

The second student replies,

> What do you mean I am not a scribe like you? When you write a document it makes no sense. When you write a letter it is illegible. You go to divide up an estate, but are unable to divide up an estate, for when you go to survey the field, you can't hold the measuring line . . . you don't know how to arbitrate between the contesting parties; you aggravate the struggle between the brothers. You are one of the most incompetent of tablet writers. What are you fit for? Can anyone say?

The first replies to each of these charges and then goes on the attack again.

> . . .you are the laziest of scribes, the most careless of men. When you do multiplication, it is full of mistakes . . . In computing areas you confuse length with width. Squares, triangles, circles, and sectors—you treat them all without understanding . . .

To this, the second responds by lauding his own keeping of accounts, by claiming that he was "raised on Sumerian" and by attacking the first's ability to write.[7]

The rather limited character of Sumerian learning implied by this essay is representative of the earliest schools and of the level of knowledge sought by most scribes then and later; but even within this constricted curriculum there is a very significant scientific component imposed largely by the structure of the language and pedagogical necessity. In order to carry out the first and most fundamental function of schooling—to teach the Sumerian language, and later both Sumerian and Akkadian—the scribal teachers devised a system in which they divided the language into groups of related words and phrases, which students copied and memorized until they could be reproduced rapidly and legibly. But because of the agglutinative nature of the language, this pedagogical technique was at the same time a method of devising taxonomic schemes and teaching a vast amount of botanical, zoological, geographical, and mineralogical knowledge.[8] Among the word lists are numerous compendia of the names of trees, insects, birds, stones, minerals, and so on. In each case, the name of a specific entity is composed of a generic root word, like tree, herb or stone, accompanied by qualifying words that indicate distinctive physical characteristics such as color, texture, and hardness. As time went on and more and more objects were known, the modifiers even came to include terms that indicated chemical reactivity; thus there are certain minerals whose names contain a term indicating that they effervesce when mixed with acids.[9]

Because of the character of their language and education, then, Mesopotamian scribes were driven, almost without intent, into a study of the properties of natural objects and the relationships between them.

Even though Mesopotamian taxonomic systems became one of the bases for later, self-consciously scientific studies of such things as mineralogy and botany, it is worth asking whether it makes sense to talk about the activities of scribes who compiled Sumerian and Akkadian word lists as scientific. To some extent all languages embody an ordering and structuring of experience; so the aims of language share with those of science the tendency toward order. Such a minimal sharing of goals is surely insufficient to argue that the creation of language is generally scientific in any important way. On the other hand, the specific activities of Sumerian teacher-scholars in forming word lists involved far more than an implicit ordering of the world. In the first place, it involved a self-conscious attempt to organize words in order to effect an intellectual economy and to make the transmission of knowledge easier. In their very self-awareness of the need to order for effective utilization and recall of information the Sumerian scribes embodied an important habit of mind associated with science. But perhaps even more importantly, the Sumerian language itself structured experience predominantly in phenomenal terms—that is, in terms of sensory information rather than in terms of affective relationships and uses. A list of thirty-one distinct kinds of sheep from Uruk, for example, included such appellations as "long-haired" and "fat-tailed" rather than "happy" and "sad," or "stewing" and "frying."[10] Thus, the language focused subsequent concern on the careful observation of phenomena, which is often a central activity of scientists. We see nothing among the works of these early scribes, however, to indicate any special concern for what we have called "underlying form." If such a concern arose in Mesopotamia it was only in connection with the relatively late (i.e., Babylonian) development of mathematical and astronomical knowledge.

In the next two sections I discuss the appearance and growth of mathematics and astronomy within the Mesopotamian school tradition, but some qualification of what I have stressed so far is necessary here. The primary mission of the Mesopotamian schools remained the training of scribes for both the sacred and secular bureaucracy, and most scribes became what we might call "junior" scribes who sought little additional learning. But the curriculum of the schools widened as time went on, especially to include a substantial "literary" component through which the myths, epics, and proverbs of the Sumerians were passed on to subsequent Mesopotamian peoples. Moreover, some scribes went on to become "scholar-teachers" (not unlike today's university professors) whose aims were to advance learning as well as to transmit it. It is to this relatively small group of men that we owe not only the creation of a tradition of belles lettres (which had immense impact upon Hebrew literature, including the Bible, and upon early Greek literature, especially that of Homer and Hesiod),[11] but also the growth of mathematical and astronomical knowledge during the second and first millennia.

THE DEVELOPMENT OF MATHEMATICS IN THE ANCIENT NEAR EAST

There are three classical texts that seek to explain the origins of mathematics in antiquity. All three were written by Greeks long after the process they were seeking to understand. Each tells us more about the attitudes of its author than about the origins of mathematics, and two of them place the occurrence in Egypt rather than in Sumer, where it almost certainly belongs. But two, at least, mention Mesopotamian mathematics. Taken together these texts help to provide a framework for understanding the emergence and development of mathematics throughout the ancient Near East.

The first Greek to write on the history of mathematics was Herodotus (ca. 450 B.C.). According to his account, the Egyptian king, Sesostris (ca. 1300 B.C.),

> . . . divided the country among all the Egyptians by giving each an equal square parcel of land, and made this his source of revenue, appointing the payment of a yearly tax. And any man who was robbed by the river of a part of his land would come to Sesostris and declare what had befallen him; then the king would send men to look into it and measure the space by which the land was diminished, so that thereafter it should pay the appointed tax in proportion to the loss. From this, to my thinking, geometry was invented and passed afterward into Greece; but knowledge of the sundial and the gnomon and the twelve divisions of the day came into Greece from Babylon.[12]

Two things about this account are worth noting. First, the origin of mathematics is attributed to a practical social need—that of measuring land for tax purposes. Second, though Herodotus attributes geometry to Egypt he shows an awareness that other knowledge related to mathematics came from Babylon, a direct inheritor of Sumerian culture. On both of these points Herodotus is fundamentally correct. Though the earliest mathematical texts in both Egypt and Mesopotamia predate the reign of Sesostris by at least a millennium, all of the early evidence indicates that mathematics originated as a response to practical needs. This fact is embedded in the very term "geometry," which means "land measuring." The great bulk of Egyptian mathematical knowledge, for example, is expressed in the form of problems faced by people like engineers, quartermasters, bakers, or brewers. In a school text of about 1700 B.C., which is reminiscent of the Sumerian disputation over the relative abilities of two students, one Egyptian scribe chides another for his inability to solve some typical problems.

> You are given a lake to dig. You come to me to inquire concerning the rations for the soldiers, and you say, "Reckon it out." You are deserting your office, and the task of teaching you to perform it falls on my shoulders . . . I cause you to be abashed when I disclose to you a command of your lord, you who are his royal scribe . . . A ramp is to be

constructed 730 cubits long, 55 cubits wide, containing 120 compart-
ments, and filled with reeds and beams; 60 cubits high at its summit,
30 cubits in the middle, with a batter of twice 15 cubits and its pave-
ment 5 cubits. The quantity of bricks [assuming a standard size]
needed for it is asked of the generals, and the scribes are all asked
together without one of them knowing anything. They all put their trust
in you and say . . . answer how many bricks are needed for it?

See, its measurements are before you. Each one of the compart-
ments is 30 cubits and is 7 cubits broad.[13]

In Sumer, as we have already seen, the earliest mathematical
computations appeared, together with the earliest writing, around
3500 B.C., and were very clearly produced for practical purposes.
Without such inventory lists the complex economic dealings of the
temples would have been impossible to manage. Before 3000 B.C.
these inventories were joined by deeds of sale that stipulated unit
prices, numbers of units purchased, and total costs, and by contracts
to borrow seed grain at a stipulated annual interest, which showed
the total repayment amount of principal plus interest. Thus, there
can be no doubt at all that Sumerian mathematics, like writing,
originated as a utilitarian aid to commercial activity. Even the
Sumerian number system, which was developed during the third
millennium, reflects this origin, for it is a place value system like our
own decimal system, but based on multiples of 60 rather than 10. In
this sexagesimal base it followed the traditional system of measures
of weight in which 60 shekels make one mina and 60 mina make one
talent (i.e., where each large unit was 60 times the immediately
smaller one). Thus, the number system was specially designed for
convenience in commercial transactions. That such a sexagesimal
place value system is also particularly convenient for doing virtually
all arithmetic calculations is a happy circumstance that the Egyptian
non-place-value system of notation missed. When the Mesopotamian inheritors of the Sumerian number system began angular
measurements and measurements of time for astronomical purposes
in the late third millennium they naturally produced the 60-second
minutes, 60-minute hours, and 60-minute degrees which were transmitted into Greek and hence into European usage.

All of the early evidence, then, tends to support Herodotus in
his assertion that mathematics had a practical origin and that astronomical measures came into Greece from Babylon. On the face of it
this evidence completely discredits Aristotle's alternative account of

the origins of mathematics. For Aristotle argued that mathematics was a nonpractical activity that could exist only with the growth of a leisure class.

> Hence when all such inventions [i.e., those produced for the necessities and conveniences of society] were already established, the sciences which deal neither with the necessities nor the enjoyments of life were discovered. And this took place earliest in the places where men first began to have leisure. That is why the mathematical arts were founded in Egypt, for there the priestly caste was allowed to be at leisure.[14]

However, if we consider the development of mathematics during the second millennium B.C. rather than at its origins, and if we insert Mesopotamia for Egypt in Aristotle's account, there may be an extremely important element of truth in what he had to say. The evidence indicates that Mesopotamian mathematics reached great sophistication and power during old Babylonian times. It did so in the hands of a class of learned men who earned leisure for scholarly activity by teaching in the schools (which were often attached to the temples). And it did so only by turning away from the immediately practical problems of measurement and commerce to deal generally and abstractly with the properties of numbers and the manipulation of quantities. Whether there could have been a significant development of mathematics without some freedom to follow conceptual concerns beyond the demand of practical problems is uncertain. But it is clear that historically in Mesopotamia, where some scholars apparently became interested in "pure" number theory, mathematical development went far beyond that in Egypt, where there is no evidence for a strong nonutilitarian tradition.

By the end of the old Babylonian period, sometime around 1600 B.C., Mesopotamian mathematics had reached a high level of sophistication. The bulk of the problems presented in the thousands of tablets that we have from this period remain tied to utilitarian concerns—to determining the amount of materials needed to build structures of a variety of shapes and sizes; to figuring out the distribution of given quantities of commodities when individuals get differing portions depending on rank or function; to calculating the interest on loans. Other tablets provide tables of such things as multiples, reciprocals, square roots, cube roots, and squares, which are useful supplements to practical calculations. A small number of tablets, however, present problems and develop techniques whose sole function seems to have been to extend knowledge of the properties of numbers without any thought of application.

The evidence for a theoretical element in Mesopotamian mathematics is somewhat indirect, for the Sumerians and Babylonians never developed the habit of using arbitrary symbols for unknown quantities. All problems were presented using concrete numbers and no general symbolic equations were written as shorthand for whole classes of equations. Nonetheless, there can be no question, for example, that the Babylonians knew the general method of generating what we call Pythagorean triples; that is numbers a, b, and c, such that $a^2 + b^2 = c^2$.[15] Tables of such triples, along with supplementary variables used in modern techniques of generation, appear from about 2000 B.C. Some Pythagorean triples might well have been useful for determining right angles and in land measuring, but the tables go beyond triples of any conceivable practical value. An additional clear example of the Babylonian texts' going beyond practical applications appears in the following problem, which is of a common type done from about 1800 B.C.[16]

> Length, width, I have multiplied length and width, thus obtaining the area. Then I added to the area, the excess of length over width: 3,3 (i.e., 183 was the result). Moreover, I have added length and width: 27. required; length, width, and area.

There are simply no practical problems in which linear dimensions are added to areas. The above problem is simply a way of expressing what we would write as the system of equations,

$$xy + x - y = 183$$
$$x + y = 27.$$

The Babylonian tablets contain many such problems, which all seem to be reduced to standard forms and solved by the application of rules that are used without ever being generally stated. Only rarely are such problems of any conceivable applicability to practical life.

The pattern of development shown by Mesopotamian mathematics—one in which a field of intellectual concern arises for utilitarian purposes that it may subsequently transcend—is a common one in connection with the sciences. This pattern was nicely recognized by the third Greek interpreter of ancient mathematics, Proclus Diadochus (ca. A.D. 450), who agreed with Aristotle that pure knowledge was somehow more dignified than utilitarian knowledge, but who also recognized the accuracy of Herodotus' account of the applicate origins of geometry. Proclus reconciles the two earlier accounts.

> It was, we say, among the Egyptians that geometry is generally held to have been discovered. It owed its discovery to the practice of land measurement. For the Egyptians had to perform such measurements because the overflow of the Nile would cause the boundary of each person's land to disappear. Furthermore, it should occasion no surprise that the discovery both of this science and of other sciences proceeded from utility, since everything that is in the process of becoming advances from the imperfect to the perfect. The progress, then, from sense perception to reason to understanding is a natural one. And so, just as the accurate knowledge of numbers originated with the Phoenicians [read Mesopotamians] through their commerce and their business transactions, so geometry was discovered by the Egyptians for the reason we have indicated.[17]

I have taken substantial space to look at the origins and growth of mathematical knowledge in the ancient world for two basic reasons. First, in connection with mathematics, evidence first clearly indicates the existence of what Poincaré (see chap. 1) suggested was an important aspect of the scientific mentality—that is, the search for universal underlying patterns of order that may go beyond the immediate and apparent needs of men. This tendency is clearly evidenced in Babylonian mathematics by the middle of the second millennium; and even the greatest Egyptian mathematics texts, the Rhind papyrus, which dates from about 1700 B.C., hints at the tendency in its introduction, which claims to offer "Rules for enquiring into nature, and for knowing all that exists, every mystery, . . . every secret."[18]

Second, and of more immediate significance, mathematical learning is a prerequisite not only for many practical activities, but also for numerous other sciences. In particular, it underlies astronomy, which was the science that had by far the most important direct impact upon other aspects of ancient culture. So in order to understand an important part of the context out of which scientific activities and habits of thought began to extend their impact upon ancient life, one must be aware of the strong and sophisticated Mesopotamian mathematical tradition.

SUMERIAN RELIGION AND THE RISE OF ASTRONOMY

Unlike Christianity and Islam, the dominant religions of later Western cultures, the religions of Sumer and later Mesopotamia were not based on the interpretation of a single or a small number of canonical texts. There is thus no tradition of exigetical (i.e., explanatory and interpretive) literature from which we can easily learn or infer the specific religious dogmas, credos, or beliefs of the ancient Near East. Instead we must piece together our understanding of the religious life of the Sumerians and the inheritors of their culture from architectural remains, the artifacts associated with religious ritual, fragments of a small number of hymns to the gods, and from myths, legends, and even portions of ancient tablet catalogs, whose primary purposes were often secular rather than religious. The reconstruction of religious beliefs and practices has been a complicated task involving substantial speculation, which is far too complex to consider in detail here. Nonetheless, it is crucial for us to selectively summarize what is known or generally thought about the religious dimensions of Mesopotamian life for two related reasons.

In the first place, religious practices and beliefs provided the most universal and integrative elements of Near Eastern culture. In a society characterized by economic specialization and class distinctions, religion provided common experiences and conceptions. As far as we know, all Sumerians, whether men or women, rich or poor, even free or slave, worshipped the same gods in much the same way, and shared a wide range of beliefs about the gods and their relation to humans. Furthermore, religion was not only universal in culture, it was undeniably and universally important. This importance is reflected in Mesopotamian architecture and in the energy, skill, and wealth that went into producing devotional artifacts. It is reflected in political documents, and above all is reflected in the fact that religious rituals accompanied and (to some extent) determined the circumstances (especially the timing) of major political and economic activities. For this reason, any developments related to religious belief and practice were potentially important to nearly all aspects of Mesopotamian culture.

The second reason for our special interest in religion derives from the centrality of religion in Mesopotamian culture and the consequent capacity for religious practices to impinge upon a vast range of activities. Sumerian religion, from its inception, involved a significant, though not a dominant, concern with the heavens. Three of the seven most important Sumerian deities were associated with the three most prominent heavenly bodies: the sun, the moon, and Venus. So there was encouragement for astronomical observation within the religious tradition. Furthermore, Mesopotamian religion involved an element of omen interpretation and prognostication, or divination, very early. This interest was neither dominant nor primarily concerned with celestial omens at an early date, but it encouraged some interest in celestial events and their possible portent. These concerns with celestial phenomena within Sumerian religion gave rise to an initially very modest tradition of astronomical observation and record keeping among the scholars associated with schools and temples by the late third or early second millennium. Slowly, this tradition began to incorporate mathematical techniques developed within the schools, becoming more complex and sophisticated. And, as astronomical knowledge increased, celestial phenomena became increasingly important in religious practices, including omen interpretation. Finally, as we shall see in the next section, astronomical knowledge became so important that it not only permeated Babylonian religion, it transformed it. All other forms of divination gave way to astrology, which was to remain a powerful cultural force in the Hellenistic world and within Islam and Christendom, at least until the late Renaissance. Even Mesopotamian agricultural practices were modified by astronomy as it was mediated through religious institutions. Thus, Mesopotamian religion is important as the locus of the first great impact of scientific knowledge upon Western culture.

The first major point to be made about early Mesopotamian religious sensibility is that the Sumarians were polytheists who conceived all deities as personifications of natural objects, events, or phenomena. Conversely, all natural objects, events, or phenomena were manifestations of, or under the direct control of gods. Nothing in the natural world, then, was without a living and personal aspect. Henri Frankfort expressed this characteristic by saying that men lived in an I-thou relationship with natural entities.[19] All objects were capable of responding to human needs; but, like humans, they often had to be cajoled into doing what men wanted and rewarded for their benevolent acts by adulation and material rewards or offerings. Thus, for example, a sick man prayed to *salt* to make him well.

Oh. Salt, created in a clean place;
For food of Gods did Enlil destine thee.
Without thee, no meal is set out in Ekur.
Without thee, god, king, lord, and prince do not
      smell incense.
I am so-and-so, son of such-and-such,
Held captive by enchantment,
Held in fever by bewitchment.
Oh Salt, break my bewitchment, loose my spell,
Take me from my enchantment!
And as my creator
I shall extoll thee.[20]

    Of course, not all things as gods—or more properly, not all gods which manifested themselves in things—had the same stature. There was a relatively small group of fifty major gods headed by an assembly of seven powerful gods, manifested in the most powerful forces of nature. The seven powerful gods debated and decided upon the course of all things and the fates of all men, much as the later Greek Olympian pantheon gods did. There was Anu, god of the sky and initially most powerful among all gods, whose name was the ordinary word for sky; there was Enlil, literally "lord-storm," through whose force the dictates of the assembly of gods were usually enforced; there was Ninmah, or Nammu, the female god of the earth and fertility; and Enki, lord of the waters, who was the cleverest of the gods just as the designers and managers of the canal systems were the cleverest among men. Finally, there were three celestial gods: Utu, or Shamash, the sun god; Sin, or Nana, the moon god; and Inanna, or Ishtar, associated with Venus and the goddess of love.

    One can get some sense of how the Mesopotamians venerated these gods and interpreted phenomena as results of their actions by reading a portion of a hymn dedicated to the water god, Enki.

Oh Enki, master over prudent words, to thee
    I will give praise.
Anu thy father, pristine king and ruler
    over an inchoate world,
Empowered thee, in heaven and on earth, to guide
    and form,
      exalted thee to lordship over them.

To clear the pure mouths of the Tigris and Euphrates,
*   to make verdure plentiful,*
Make dense the clouds, grant water in abundance
*   to all ploughlands,*
To make corn lift its head in furrows and make
*   pasture abundant in the desert,*
To make young saplings in plantations and in orchards
*   sprout, where planted, like a forest—*
These acts did Anu, king of gods, entrust to thee . . .[21]

Considering the nature of their gods, we can speculate with some assurance on the importance and function of at least some Sumerian religious rituals. By early in the second millenium, in almost every city just prior to the spring floods of the Tigris and Euphrates rivers, the major religious festival of the year occurred. In this new year's festival the central feature was a ritual marriage of the gods Ninmah and Enki—of the earth and the water. This marriage was enacted by the king and a selected priestess, accompanied by votive offerings to the two and by prayers for the water god to return from his exile from the land and for the fecundity of the earth goddess. By exhorting the gods, symbolically reenacting their union, and by making offerings to them, the people clearly hoped to influence the gods and bring about Enki's reappearance so that the fertile earth could be impregnated and bring forth the next year's crops. These rituals were seen as crucial to the success of the most important practical activities of the society. To fail to satisfy the gods was to invite disaster on a grand scale for the entire community.

Because most of the Mesopotamian's day-to-day concerns were with agricultural, commercial, and political events, it should not be surprising that the bulk of surviving hymns, myths, and epics should deal with the gods of earth, water, and storms or atmosphere (i.e., those whose manifestations impinge most directly upon human activities). So, in its early stages, Sumerian religion was only secondarily concerned with the celestial deities. Two circumstances, however, focused some attention on the heavens. The first and simplest of these circumstances has to do with the calendar and time-keeping. No agrarian people can ignore the seasonal variations of the intensity of the sun's rays and consequent variations in temperature, nor the seasonal variations in rainfall and flooding. And a commercial people need some well-regulated calendar so that dates for the repayment of debts can be fixed. In Sumer the calendar was controlled principally by the lunar cycle; and like every significant cultural feature this fact had somehow to be acknowledged and accounted for. Thus, in the greatest Mesopotamian cosmological hymn or poem, the *Enuma Elish*, which dates from about 1800 B.C. (after the political center of Mesopotamia had moved to Babylon, and Marduk, the principle city god of Babylon, had been made head of the pantheon), we find the following passage, which has important astronomical content:

Then Marduk created places for the great gods.
He set up their likenesses in the constellations.
He fixed the year and defined its divisions;
Setting up three constellations for each of the
          twelve months
When he had defined the days of the year by the
          constellations,
He set up the station of the zodiac band as a measure
          of them all,
That none might be too long or short . . .
He caused the Moon to shine forth; and put the night
          under her command
He appointed her to dwell in the night and to mark out
          the time;
Month after month unceasingly he caused her disk to grow.
At the beginning of the month, as thou riseth over the
          land,
Thou shalt shine as a horned crescent for six days;
And with half a disk on the seventh day.
At the full moon thou shalt stand in opposition to
          the Sun, in the middle of each month.
When the sun has overtaken thee on the eastern horizon,
Thou shalt shrink and shape the crescent backwards.
As invisibility approaches, draw near to the path of
          the sun.
And on the twenty-ninth day thou shalt stand in line with
          the sun a second time.[22]

We shall return to the calendarial astronomy reflected in this passage and to its connection with agriculturally associated religious ritual in the next section; but first I shall discuss the second and more directly religious motivation for early astronomical science.

As already stated, the Mesopotamian pantheon gods decided the course of events and fates of men. At least some of these decisions were made well in advance of the events, whether natural or human, to which they applied. The Sumerians, not unnaturally, hoped that if they could not influence the gods to decide favorably for them through appropriate rituals, they might at least get some foreknowledge of events so as to be able to prepare for them properly. A very similar hope is illustrated in the famous Chinese "oracle bones" that are associated with the first known Chinese writing. In order to obtain knowledge of the future, the Sumerian priests, like the Chinese, examined and interpreted omens. At first divination was done largely through the inspection of the entrails of sacrificial animals; numerous divining texts in the form of clay or bronze models of sheep's livers attest this practice. But during the third millennium celestial events and objects became increasingly important as omens, and astrology was born.

Why the change in omenology began to occur we do not know. Possibly the growth of calendrical astronomy led to a realization that seasonal changes, and thus agricultural events, were signaled by the appearance of certain stars, and this stimulated interest in celestial signs and portents. For whatever reason, numerous documents from 2500 B.C. onward attest a growing interest in celestial omens. In an inscription from about 2500 B.C., for example, King Gudea of Lagash is described as having a dream in which a goddess appeared to him.

> She held the shining stylus in her hand, she carried a table with favorable celestial signs and was thinking; . . . she announced the favorable star for building the temple.[23]

In a famous myth from about 2300 B.C., which recounts the origins of shade tree gardening, the following introductory lines appear:

He lifted his eyes towards the lands below,
*   Looked up at the stars in the east,*
Lifted his eyes toward the lands above,
*   Gazed at the auspicious inscribed heaven,*

From the inscribed heaven learned the omens,
   Saw there how to carry out the divine laws,
Studied the decrees of the gods . . .[24]

A recently discovered catalog of tablets from about 2200 B.C. even seems to imply that eclipses may have been used as omens at this date, for one tablet is described as beginning with "Who knows the eclipses, the mother of him who knows the incantations."[25]

   By the beginning of the Old Babylonian period (ca. 1800 B.C.), though divination by inspection of entrails was still the dominant practice, the celestial aspect of omen interpretation had become so important that all other forms of divination were seen as being under the control of celestial deities. This can be seen from the following beautiful hymn:

They are laying down, the great ones.
The bolts are fallen; the fastenings are placed.
The crowds and people are quiet.
The gods of the land and the goddesses of the land,
Shamash, Sin, Adad, and Ishtar,
Have betaken themselves to sleep in heaven.
They are not pronouncing judgement,
They are not deciding things.
Veiled is the night.
The temple and the most holy places are quiet and dark.
The traveller calls on (his) god;
and the litigant is tarrying in sleep.
The judge of the truth, the father of the fatherless,
Shamash, has betaken himself to his chamber.
Oh great ones, gods of the night, . . .
Oh bow (star) and yoke (star),
Oh Pleiades, Orion, and the dragon,
Oh Ursa major, goat (star), and the bison,
stand by, and then,
In the divination which I am making,
In the lamb which I am offering,
Put truth for me.[26]

   As B. L. Van der Waerden has said, "Here it is apparent that the very ancient art of hauruspicy (entrail divination) is being placed by the poet under the tutelage of the celestial gods. No longer is knowledge of the future to come from the sacrificial lamb itself, nor even from the particular god to whom it is offered, but from the stars, called upon to witness and implant the truth."[27]

From the same period comes also the first known list of astrological prognostications, all based on observing the moon and sky on the first evening of the new year.

> If the sky is dark, the year will be bad.
>
> If the face of the sky is bright when the New Moon appears and (it is greeted with joy), the year will be good.
>
> If the North Wind blows across the face of the sky before the New Moon, the corn will grow abundantly.
>
> If on the day of the crescent the Moon God does not disappear quickly enough from the sky, "quaking" [probably some disease] will come upon the land.[28]

Like virtually all Near Eastern astrological predictions until almost Hellenistic times (ca. 300 B.C.), these simple and primitive prognoses deal not with the fate of individuals (kings excepted) but with such things as the welfare of the country and the nature of crops.

Given the growing importance of celestial divination in Old Babylonian culture we might expect to discover that astronomical observations were being made, and that scholars were becoming more interested in astronomy. That both of these things did in fact occur, and that they were directly connected with astrological concerns, is proven by the content of the earliest actual astronomical text known from Mesopotamia.

This text, the content of which can be dated to within a few years of 1561 B.C., contains both an observational section and a theoretical one, which is especially interesting as the first instance of a fusion between celestial observations and mathematical learning that was later to become extremely impressive and important. The observational section of the text records virtually all heliacal risings and settings of Venus over a period of twenty-one years, from 1581 B.C. to 1561 B.C. (the heliacal rising of a star occurs when it just appears over the horizon before the sun rises and obliterates it.) Accompanying each observation is an associated forecast. For example,

> if, on the 15th day of the month Shabatu, Venus disappeared in the west, remaining absent in the sky three days, and on the 18th day of the month Shabatu, Venus appeared in the east, [there will be] catastrophies of kings; Adad will bring rains, Ea subterranean water; king will send greeting to king.

The prognostications of this text are substantially more complex than those seen in the earlier astrological tablet; and the peculiar grammatical structure, which moves from past to future tense, strongly suggests the nature of the protoscientific underpinnings of astrological prognostication. If it is true that the celestial gods control certain events by their actions, and if it is also true that the celestial gods are repetitive in their behavior (i.e., that the motion of the heavenly bodies is periodic, or cyclical), then it ought to be true that if a certain event followed once from a certain behavior of a heavenly god, it might well occur again when the god's behavior repeats. Thus, the observation-forecasts of this Venus text can be seen as shorthand expressions for the following sort of reasoning: During a certain year Venus was observed to have its last observable heliacal rising on Shabatu 15. Three days later, on Shabatu 18, Venus appeared in the sky on the opposite side of the sun just as the sun set. Moreover, shortly after these events, or nearly simultaneous with them, a number of significant terrestrial occurrences were noted: something terrible happened to the king; it started to rain; further, there were high tides, the river rose, or a spring was found; and some letters arrived at the local royal court from another king. Then whenever Venus has its last appearance as the morning star on Shabatu 15 and its first appearance as the evening star on Shabatu 18 we can expect the same natural and political events to be repeated.

Some notion of the regularity of causal sequences as well as a recognition that the heavenly bodies periodically repeat all of the significant patterns of behavior associated with them are thus implicit in the observational section of the Venus tablets.

The theoretical sections of the tablets exploit the periodic nature of celestial motions in a new and different way—one that is so important that it demands very careful consideration. The passages of this section have the following slightly different form:

> If in the month Nisannu on the second day Venus rose in the east, there will be need in the land. Until the sixth Kislimu she will stay in the east, on the seventh Kislimu she will disappear. Three months she remains out of the sky. On the seventh Adaru will Venus appear again in the West, and one king will declare hostilities against the other.[29]

Not only are the astrological consequences of a sequence of celestial events predicted here, but the sequence of astronomical events is *predicted* rather than observed. Throughout this portion of the text, in every case a period of visibility lasting eight months and five days is followed by periods of invisibility that alternate between seven days and three months. These periods simplify the true situation. In particular, the three-months invisibility period is somewhat too long. But the techniques of the tablet demonstrate a remarkable characteristic of astrological prognostication that gradually came to make it centrally important to Mesopotamian religious life as increasingly accurate theoretical astronomical schemes were devised. The astrologer, unlike any other diviner, could thenceforth not only *interpret* omens, he could *predict* their occurrence in the future. It is hard to imagine how awe-inspiring this new ability must have seemed to a people who were, by our standards, eminently superstitious. The astrologer-astronomers were able to open up a virtually limitless perspective into the future, and as a consequence, they, their art, and the science that underlay it acquired tremendous prestige and power.

The technical development of Mesopotamian astronomy, which led to the prediction of lunar eclipses by the seventh century B.C., and to a lunar theory that could foretell the exact day of first visibility of the new moon each month for literally centuries in advance by the fourth century B.C., is beyond the scope of this book. Interested readers should consult Otto Neugebaur, *The Exact Sciences in Antiquity*, and B. L. Van der Waerden, *Science Awakening*, Volume II. Our concern in the next section is to assess some of the impacts of growing astronomical knowledge upon Mesopotamian culture.

TRANSFORMATION OF MESOPOTAMIAN RELIGION, CALENDRICAL ASTRONOMY, AND ASTROLOGY

In response to the growing predictive power and astrological import of astronomy there was a radical transformation of the underpinnings of Near Eastern religious practice. Traditional religious practices were retained, but their old meanings were undermined or changed. Morris Jastrow summarized this development very nicely at the beginning of this century.

> The Star-worship which developed in Babylon and Assyria in connection with the science of the observation of the heavens was at bottom a new religion, the victory of which brought about the decadence of the

old popular beliefs. In point of fact, in the ritual of worship, in ceremonies of incantation and purification, in hymns and prayers, in the
chants of ceremonial lamentation, in old festivals in honor of the gods
of nature, just as in heptoscopy (examination of the livers of victims)
and in the other kinds of divination which were maintained up to the
end of the Babylonian empire, popular ideas always survived. The
priests would have been careful not to destroy or imperil the dominion
which they exercised over the multitude by changing the forms of
worship in the direction of the new religion. But astral doctrines could
not, for all that, fail to make their influence felt as a dissolvent force.[30]

To follow the development of astral religion in detail through
the great variety of cults produced as older Babylonian ideas were
synthesized with traditions brought in by successive invasions of
Assyrians, Persians, and Hellenistic Greeks would be impossible
here. So we shall concentrate on a few major points that had lasting
impact on Western culture.

The first, and in a way simplest, transformation involved the
virtual disappearance of all but a few of the mightiest natural gods
and their reinterpretation as celestial deities. In Sumer and Babylon
until sometime around the sixteenth century B.C. only three major
gods and some minor ones were celestial gods. This is reflected in
the fact that astrological predictions were at this time based only on
lunar and Venus phenomena, and by the fact that in the poem
dealing with the role of the stars in guiding entrail divination the
stars were designated as mere gods of the night and distinguished
from the gods that "pronounce judgements" and "decide things."
That is, in this early stage of astral concern, while the stars may have
been able to communicate the fates they did not determine them.
During the period between about 1600 B.C. and 1000 B.C., however, all
major gods became associated with celestial bodies. Marduk, the
head of the Babylonian pantheon, became associated with Jupiter.
Nergal, the god of war, became associated with Mars, and so on.
Consequently, omen texts began to relate the phenomena of Mars to
predictions about military success and failure and the phenomena of
Jupiter to the political and economic fortunes of the kings of Babylon.
Even minor nature gods were associated with stars or clusters of
stars. So the stars and planets were gradually seen more and more as
not only the messengers of the fates but also as the determiners of the
fates.

Before turning to the impact of astronomical knowledge and an
increasingly astral focus on fundamental patterns of religious belief,

I will briefly consider the way in which Mesopotamian astronomy was related to agricultural practice through the regulation of the religious calendar. This process is particularly interesting because it shows, in a very simple way, how theoretical considerations can inadvertently come to override practical considerations and produce a feeling of alienation between scientific scholarship and everyday life. It thus provides an easily seen parallel to the much more complex developments in astrological doctrines, which gradually engendered a similar alienation.

As far back as records exist, Mesopotamian calendars were luni-solar in character; that is, months determined by the synodic period of the moon (a synodic period is the time taken for a heavenly body to return to the place from which it started in relation to the sun) were somehow fit into years determined by the period of the solar agricultural cycle. As the creation epic quoted above showed, the moon was considered to be the primary time counter. Each month began at sunset of the evening when the new moon was first visible after its conjunction with the sun. Since the synodic period of the moon is 29.53059 days, this means that most months were 29 or 30 days long. But since in the early calendar the moon actually had to be observed to start the new month, if observing conditions were bad when the new moon would otherwise have been seen, a month might become 31, or even 32, days long. In such a case the following month might become only 28, or even 27, days long.

Now, since the solar or agricultural cycle is about 365.25 days long, 12 lunar months is less than an agricultural year and 13 lunar months is more than an agricultural year. This would have caused no special problem except that from preliterate times most Sumerian months had been specifically designated as related to certain agricultural and cognate religious activities: for example, the twelfth month of the year was the "month of the barley harvest," another month was "the month of the plow," and yet another "the month of planting seed corn." To each of these months belonged the seasons. In order to keep the lunar religious calendar in tune with the agricultural year, an elegantly simple technique was devised. Ordinary years consisted of twelve lunar months; but when "the month of barley harvest" began to come much before the barley was ready, an extra month was inserted, or intercalated. If, for example, the crop was not almost ready to be harvested at the end of the eleventh month, a priest would inspect the fields and declare that an additional "pre-barley harvest," should be added to the current year.

Thus, the calendar was adjusted empirically according to agricultural needs. This meant that the calendar could be modified when late floods caused a delay in planting or when a particularly hot growing season caused the grain to mature early.

From the purely agricultural standpoint, no better calendar has ever existed, and any change would necessarily produce a practically inferior agrarian calendar. As astronomy advanced and astral elements became important within Mesopotamian religion, however, the old calendar was changed from an agriculturally determined one to an astronomically determined one. Just why this occurred must be somewhat conjectural, but it seems to me that A. Pannekoek was correct when he wrote that

> the agricultural festivities, like all great and important social happenings, were at the same time religious ceremonies. . . What was necessary or adequate socially became a commandment of the Gods, strictly fixed in the rites. What by nature took place at a determined season, e.g., a harvest home, as a religious celebration was fixed at a certain date, e.g., an aspect of the moon. The service of the Gods did not allow of any carelessness; it demanded an exact observance of the ritual. The calendar was essentially the chronological order of the ritual.[31]

Whatever the justice of Pannekoek's statement, it is certain that by about 1000 B.C. the calendar had in fact become regulated by the stars, and that a much more rigid ordering of religious ritual was established. One religious menology (a text that establishes instructions for the religious practices for every day of the year) from this time sets the requirement that henceforth only the month Nisan may be preceded by an intercalary month;[32] an associated text from the same period tells us that "Star Dil-gan appears in the month Nisan; when the star stays away the month must."[33] Here we have an instruction for the intercalation of a pre-Nisan on an astronomical rather than an agricultural basis.

Although the calendar was astronomically regulated by sometime around 1000 B.C., there were still important empirical elements that made it impossible to fix a long-term sequence of religious and agricultural activities far in advance. Intercalation still depended upon actually observing the appearance of a certain star, and the beginning of each new month continued to be determined by observing the appearance of the new moon. During Persian times (ca. 600–300 B.C.), however, astronomical science was so far advanced that a totally theoretical calendar was produced that could be fixed for centuries ahead.

Sometime around 480 B.C. the so-called "Metonic Cycle" of 19 years was established as the basis for intercalation. Since there are very nearly 235 lunar months in every 19 solar years, it was decided that there should be 12 ordinary 12-month years and 7 full 13-month years in every 19-year period, and a formula for deciding which months should be intercalated was established. Thus, the calendar began to shed its empirical nature, and to gain a theoretical aspect.[34] Moreover, this theoretical element involved a long-term period relation which was not obvious to any but a learned group of astronomers who had come to recognize it only by extended study of astronomical phenomena.

Within little more than one century after this change even the complex problem of predicting the first day of visibility of the new moon for successive months without recourse to observation was solved;[35] the Near Eastern priestly desire for a calendar that could be established for at least a century in advance without any deviation from the astronomical elements to which it was tied had been satisfied. But this was accomplished only at a price, for it undermined the old fundamental and rational connections between agricultural practice and religious ritual.

The farmer quite reasonably continued to plant when the soil was ready and to reap when the grain was ripe. But he was told to celebrate these activities only when the religious calendar told him to do so; and he could no longer understand the esoteric grounds of formal religion. The official religion which had once formed the most powerful unifying institution in society had gradually become the domain of a group of learned experts and had lost its intimate connection with the everyday activities of the vast bulk of men. As Hans Jonas describes the situation, "In a one-sided development of its original astral features, the older cult was transformed into an abstract doctrine . . ."[36] As such, Jonas implies, astral religion became estranged from the bulk of the society it was supposed to serve. This view seems supported by the circumstances surrounding the development of an increasingly rigid and theoretically structured religious calendar and its separation from agricultural practice. But there were vastly more dramatic and alienating consequences of the astronomical underpinnings of Babylonian religion that bore directly on the central content of religious belief, not only in Mesopotamia but ultimately throughout the entire Hellenistic world.

One of the first pervasive results of the increasing astral focus of Mesopotamian religion was that the immediate and personal relationships between men and natural entities began to disappear.

After about 1000 B.C., though direct appeals to nature gods remained a formal feature of some religious rituals and medical incantations, the focus of religious concern changed. As nature came to be seen as ruled principally by celestial deities whose paths were unvarying and repetitive—in no way subject to the will of human beings—the Sumerian emphasis on preestablished fate (which had played a subordinate role in earlier beliefs) became ever more important, manifesting itself as an emphasis on astral determinism, which became increasingly rigid and fatalistic as astronomical knowledge became increasingly more sophisticated. As a result, not only astrology, or celestial omenology, but all forms of divination increased in importance with respect to rituals that were intended to modify the behavior of the gods toward some human end.

Until at least around the sixth or fifth centuries B.C. there is reason to believe that few Babylonian astral priests believed that the influence of the stars was completely determinative. Particular astral configurations were seen as favorable or unfavorable to specific human undertakings;[37] but within relatively broad limits men could overcome the influences of the stars through religious ritual, special personal effort, or even by choosing freely to act at auspicious rather than inauspicious times. In connection with an intensified Babylonian tradition of observational astronomy that began around 650 B.C. and lasted into the Christian era, however, the grounds for a more fatalistic astrology, which culminated in horoscopic astrology, was established.

From 650 B.C. there was a continuing tradition of daily observations of major celestial events. The texts of this tradition located planetary phenomena in the newly created signs of the zodiac, and correlated the observed celestial phenomena with weather patterns, political events, and even with fluctuations in commodity prices.[38] Thus, the data contained in these observation texts provided grounds for the growth of zodiacal astronomy and astrology, for predictive theories of the motions of all of the planets (theories necessary for the birth of a detailed horoscopic astrology), and a presumed empirical basis for assuming a rigid correlation and correspondence between celestial and terrestrial phenomena.

The doctrine that resulted has been succinctly, passionately, and (I think) correctly summarized by W. W. Tarn in the following way:

> The stars, and above all, the planets, obviously moved in the vault of heaven according to fixed laws; and a doctrine of "correspondence"

had arisen—this was the vital matter—according to which the heavens above and the earth beneath were the counterpart of each other, and what happened in the sidereal world was reproduced on earth. But the movements of the sidereal world were fixed; if then, there was correspondence, what happened on earth was also fixed; and men's actions too were fixed, for man was a microcosm, a little world, the counterpart of the great world or universe, and his soul was a spark of that celestial fire which glowed in the stars. From this sprang one of the most terrible doctrines which ever oppressed humanity, the Babylonian *Heimarmene* or Fate, which ruled alike stars, earth, and men; all their motions were fixed by an immutable Power, non-moral, which neither loved nor hated, but held on its course as inexorably as the planets across the firmament.[39]

Although such a doctrine was always opposed, even by some Babylonian astral priests, it seems fairly certain that sometime before about 300 B.C., when Babylonian calendarial astronomy had reached its theoretical apex and Alexander the Great's army almost simultaneously overran Babylon, a totally fatalistic attitude had come to the ascendant. This fact is illustrated by a marvelous story told by the Roman historian Appian about the founding of the Hellenistic city of Selucia, near Babylon, by Alexander's General Seleucus Nicandor in 301 or 300 B.C. According to Appian, Seleucus asked the Babylonian priests to determine the best hour to begin constructing his new city, but the priests, fearing (rightly) that the new town would overshadow their own, decided to trick Seleucus by naming an unlucky hour. At this point, however, predetermined fate took over. Because Seleucus was destined to found a prosperous city, the workmen inexplicably began work before the assigned hour,

> . . .the heralds who tried to stop them were not able to do so. . . . Seleucus, being troubled in his mind, again made inquiry of the *magi* concerning his city, and they, having first secured a promise of impunity, replied, "that which is fated, o king, . . .neither man nor city can change, for there is a fate for cities, as well as for men!"[40]

Mesopotamian culture, as it had emerged in Ancient Sumer and had been transmitted and modified in Babylon, gradually disappeared as a powerful and cohesive force in the Hellenistic world. But the doctrines of astral religion with their focus on astrological fatalism were incorporated into a variety of Persian, Greek, and even Roman religious cults and philosophical systems to have their greatest impact after their cultural roots had been severed.

The first non-Mesopotamian cult to adopt fundamental doc-

trines from Babylonian astral religion and astral fatalism was probably the cult of Zervan, a Persian religious movement that arose during the seventh or eighth century B.C., preceding the Persian conquest of Babylon but during the time when cultural contacts were growing. Zervan was the Persian God of time, and as such was naturally associated with the celestial time counters and with astronomy. According to the fullest expression of Zervanist doctrine that makes explicit the implicit assumptions of Babylonian astrological fatalism, not only does all that happens come from the stars but (because the course of the stars is predetermined) all things that happen on earth must also be inevitable. Thus, a Persian source from between A.D. 200 and 650 tells us that "All fortune, good and ill, that befalls man, comes from the twelve [signs of the zodiac] and the seven [planets]," and that, "at the appointed time, that will happen which must."[41]

While we might reasonably doubt that the extreme fatalism of this late Zervanist doctrine was present at the beginnings of the movement, there can be virtually no doubt of its source in Babylonian doctrines.

Such ideas were also intimately connected with the doctrine of the Great Year and Eternal Return which came to dominate much Greek and Roman thought through Pythagorean, Platonic, and Stoic formulations during a period that lasted from about 500 B.C. to at least A.D. 400. These doctrines have numerous and important variations, but the simplest and most basic expression is presented in Nemesius' account of Stoic fatalism.

> The Stoics explain: the planets return to the same celestial sign, where each individual planet originally stood. . ., in certain times the planets bring conflagration and annihilation of all things; then the world starts anew from the same place, and while the stars turn the same way as before, . . .everything will be the same and unchanged down to the minutest details.[42]

Although astral fatalism was almost solely a product of Babylonian astronomical knowledge developed within the context of Mesopotamian religious beliefs, it had its major impact on the Greek and Roman world in connection with a set of attitudes that were neither Mesopotamian nor Greek, but were Persian in origin. Zervanism was the first Persian cult to incorporate astral doctrines, but the much more important Persian cult of Zoroaster with its focus on

ethics, morality, and the salvation of the soul made astral doctrines not only palatable but critically important in Hellenistic culture.

Sumerian and Old Babylonian religion were not predominantly ethical in their orientation. The nature deities of Mesopotamia personified phenomena rather than abstract concepts like good and evil, and almost all of the Sumerian and Babylonian gods were morally ambivalent—capable (like the men after whom they were patterned) of being truthful or deceitful, of loving or hating, of being gentle or brutal. In addition, the major concerns of the older Mesopotamian religions were public and of this world. Though there were personal gods, they were called upon to preserve and restore health or to sustain and improve one's wealth and station in life. There was little or no concern with salvation in an after life. In fact, when men died, all supposedly descended to the underworld where they lived a life that was a sort of bland and pale reflection of the life they had lived upon earth. There was no great judgment that separated the good from the bad and sent the good to live in a separate and wonderful place while the bad were consigned to some counterpart of the later Christian hell.

In contrast with the older Near Eastern ideas, Persian doctrine, as expressed in the Avesta of Zoroastrianism (probably from the 6th century B.C.), was fundamentally ethical, dualistic, and highly personal or individualistic. In the Zoroastrian cosmos two groups of powers confront one another in ceaseless conflict. The first group, headed by Ahura Mazda, the Great God, includes Good Thought, Right Order, Holy Character, Health, and Immortality. The other group is composed of the evil counterparts of the first. Between these powers of good and evil each man stands alone. His duty is to serve the powers of good, but he may choose the evil. When men die they are judged by fire, which is the symbol of Ahura Mazda, and they travel over the Civnat Bridge from which the righteous souls *ascend* to live with Ahura Mazda, and the evil ones fall off into the underworld of hell. The ascension of the good souls is, in all early versions of this religion, through the three regions of good thought, good words, and good action to the light world of Ahura Mazda.

In early versions of this doctrine there is a small element of astral concern, for when the righteous souls pass over the abode of the good gods we are told that "the stars, moon, and sun will bless them."[43] Thus, when this Persian religion came into contact with the astral religion of the Babylonians it was natural that some synthesis should occur. The new blend of ideas retained the ethical and personal cast of the Persian ideas, but integrated astral doctrine. Thus, for example, the ascent of the soul through the three abstract realms of good thoughts, good words, and good actions became an ascent through the seven spheres of the planets; and the abode of Ahura Mazda became the sphere of the fixed stars. In fact, the abstract Persian deities were gradually concretized as astral deities. As one might expect, the greatest Persian god, Ahura Mazda, eventually took over the planet Jupiter, which had been the star of the most powerful Babylonian god, Marduk, and other appropriate associations were made.

Without a doubt, the most lasting impact made by this change was in the justification it provided for personal horoscopic astrology, which originated in Babylon between the sixth and fifth centuries B.C. while it was under Persian rule, and which spread throughout the Hellenistic world to have major importance within Christianity and Islam.

By the fifth century B.C. Mesopotamian astronomical theory was capable of determining the positions of all major heavenly bodies for any date in the past or future, so all of the technical elements upon which horoscopic astrology depend were present. Moreover, it was but a small step from thinking of the soul ascending to heaven after death through the celestial gods to thinking of the soul descending through the stars before birth and being given its attributes by the celestial deities. If one could tell the positions of all the appropriate stars and planets at a person's nativity, then one could, presumably, predict both a person's character traits and perhaps even propitious and dangerous times for him or her to take decisive actions.

The first datable horoscope is a Babylonian tablet from 409 B.C., but the practice of horoscopic astrology spread rapidly. Plato shows some knowledge of its doctrines by about 350 B.C. In the myth of Er, which constitutes the final Book of *The Republic*, he both recounts a theory of astral "necessity" linked to individuals' fates and rejects the fatalistic implications by arguing that each soul is initially free to "choose" its own fate.[44]

After the conquest of Babylon by Alexander astrology was carried throughout the Hellenistic world in the doctrines of Stoicism, which became the dominant philosophical movement of Hellenistic Greece and Rome.

Perhaps the clearest short expression of the grounds and content of Stoic astrological fatalism was written by the Roman orator

Cicero (an antifatalist) during the period when astrology was becoming an increasingly powerful and pervasive doctrine in Rome.

> In the starry belt which the Greeks call the Zodiac there is a certain
> force of such a nature that every part of that belt affects and changes the
> heavens in a different way, according to the stars that are in this or in an
> adjoining locality at a given time. This force is variously affected by
> those stars which are called "planets" or "wandering" stars. But when
> they have come into that sign of the zodiac under which someone is
> born, or into a sign having some connection or accord with the natal
> sign, they form what is called a "triangle" or "square." Now since
> through the procession and retrogression of the stars the great variety
> and change of the seasons and of temperature take place, and since the
> power of the seen produces such results as are before our eyes, they
> believe that it is not merely probable, but certain, that just as the
> temperature of the air is regulated by this celestial force, so also
> children at their birth are influenced in soul and body and by this force
> their minds, manners, disposition, physical condition, career in life
> and destinies are determined.[45]

To discuss the extended development and impact of astrological doctrines in detail would be impossible and inappropriate here, but a few comments are in order. It would be foolish to underestimate the power and influence of astrology simply because most of us now believe it to be misguided and superstitious. Astral doctrines and astrology became central tenets of virtually all major Hellenistic religious cults, whether in Egypt, the Near East, or in Greece and Rome, during the period between 300 B.C. and A.D. 300. Even cults that were primarily nonastral, like Christianity and Judaism, took on a special doctrinal shape because they had to combat the dominant astral mood. Furthermore, astrology played an important practical role in republican Rome, in medieval Islam, and in Renaissance Europe. Many political figures in the civilized Western world would not think of making an important decision without consulting their astrologers or "mathematicians." This was even true of Pope Urban VIII, who was offically opposed to astrological doctrine but who could not escape the pervasive aura of belief in celestial influences.

Finally, it should be emphasized that just as the Babylonian calendar became abstract and based on complicated astronomical knowledge that was alien to the general populace, astrology, which was initially a relatively simple doctrine, became increasingly complex and esoteric between its inception around 400 B.C. and its most prestigious codification by Claudius Ptolemy in the second century A.D. And this movement toward greater complexity made astrology something to be feared as well as admired and believed. This was especially true because the mysterious (i.e., esoteric) doctrines of astrology, which could be understood by only the few persons trained in astronomy and mathematics, were at the same time supposed to play a determinative role in the life of every human being.

THE SCIENTIFIC CHARACTER OF ANCIENT ASTROLOGY AND ASTRAL RELIGION

Because the fundamental aim of this book is to analyze the impact of scientific activities, attitudes, and ideas or theories upon a wide variety of human beliefs and behaviors, it seems proper at this point to reflect briefly on why it should seem correct to see ancient astral religion and astrology as the outgrowth of scientific activity, as embodying scientific attitudes, and as critically dependent upon scientific theories. Such reflection is particularly necessary because in the modern world we tend to draw a very clear distinction between astronomy, which we define as a science, and astrology, which most intellectuals decry as superstitious and unscientific; and because the scribes and priests who developed the knowledge underlying astral religion and the doctrines of astrology were almost certainly not aware of the distinctions that we make between scientific and unscientific activities, attitudes, and ideas.

　　In any discussion of the links between scientific activities and ideas and "other" kinds of activities and ideas we must constantly guard against the tendency to confuse "scientific" with "correct according to the latest scientific doctrines." The argument that ancient and Hellenistic astrology is inconsistent with the canons and content of twentieth-century astronomy has little or no bearing on questions of the relationship between astronomy and astrology in the ancient world.

　　Whatever their motives for considering the motions of the heavenly bodies, it is clear that the activities and the written products of Sumerian and Babylonian scribal studies of the heavens incorporated many of the characteristics of science that were discussed in chapter 1. They both sought after and discovered order, uniformity, and repetitive patterns in the heavens. These patterns were quantified, and used to provide foreknowledge or predictions of future celestial events, the appearance of which both tested and validated the period relations that had been discovered. Moreover, both the Venus tablets from the sixteenth century b.c. and the long sequence of astronomical diaries that appeared from the seventh century to the end of Babylonian culture (as well as the astrological systems derived from them) indicate an attempt to establish that the regularity of heavenly motions first announced, and later *acted as a cause of*, the underlying form or structure of terrestrial events. The knowledge of the relationships between celestial motions, and between celestial and terrestrial events, was also taken to be universally valid; this was because the period relations of the stars and planets were assumed to carry over from one place to another and to be invariable over time, and astral determinism was assumed to operate regardless of whether an individual believed or disbelieved in its efficacy. That is, faith—a subjective characteristic—was taken to be totally irrelevant to the objective influences of the stars.

Finally, we should emphasize that Mesopotamian astral knowledge was phenomenal (i.e., it focused on the sense data provided by vision with regard to the stars and upon observed events in social life, and later in the life of an individual). This phenomenal emphasis was expressed particularly in Cicero's emphasis on the "power of the seen" (see previous section). It is certainly true that as horoscopic astrology developed in the Hellenistic world it focused increasingly on affective relationships—temperament and moral qualities like lust, greed, avarice, and bravery. But the Babylonian roots for this development were equally certainly based on the observation of phenomena and the correlation of celestial and terrestrial occurrences.

Thus, in their concern with discovering order and underlying form, in their focus on phenomena, in their assumption that their knowledge was universally valid and independent of the human observer, the Mesopotamian scribes and astral priests employed fundamentally scientific habits of mind; and in their observing, quantification of period relations, prediction of celestial events, and correlation of different kinds of phenomena, they were engaged in fundamentally scientific activities. Since astral religion and astrology were undoubtedly based upon the products (the calculating schemes) of these activities and upon the habits of mind of these scribes and astral priests, they were the products of *scientific* activities and habits of mind.

ANTISCIENTIFIC THOUGHT IN THE ANCIENT NEAR EAST

If it is true that the increasingly scientific cast of Near Eastern astral religion that developed in connection with growing astronomical knowledge played a significant role in undermining the popular appeal and power of the official astral "state" religions of Babylon (and later of Persia), and that the exclusive knowledge of astral determinism produced a fearful response in some, then we might expect to find an antiastral, even an antiscientific, element among some of the many cults that sprang into existence in the Hellenistic world. In this section I argue that we do find such elements, at least in several Gnostic cults, whose origins date very roughly from the time of the origin of Christianity. My argument here must be very tentative and somewhat complicated because the early opposition to scientific thought, though not to astral doctrines, is indirect and implicit rather than explicit. Nonetheless, it seems important, in part because there has been a revival of Gnostic elements in very recent antiscientific thought,[46] and one of my aims in this book is to examine the genesis of contemporary attitudes toward science. In addition, as discussed in chapter 5, Gnostic attitudes toward the phenomenal world deeply influenced Christian doctrines.

Gnostic religions—literally religions of "knowledge"—were all, so far as we know, highly syncretistic. That is, they all borrowed from a variety of traditions, some Babylonian, some Persian, some Jewish, some Hellenic and Hellenistic Greek, and even some Christian. Furthermore, no two of the many Gnostic sects of antiquity were doctrinally identical. Thus, what I say of Gnosticism will be neither complete nor necessarily true of all sects. But it will deal with central tenets of the vast bulk of Gnostic religions.

For our purposes, the most interesting Gnostic sect is that of the *Mandeans*, which arose near the Euphrates and still survives as a minor sect in modern Iraq. Since this group arose in the region of earlier Babylonian dominance, it understandably shows the most direct response to Babylonian astral religion, although its basic attitudes were shared by nearly all Gnostic religions.

The Mandean repugnance toward the older astral religion of Babylon and Persia is demonstrated not by denying the importance of astral deities, but by identifying the stars and planets with the old Persian powers of evil rather than with the powers of good. To achieve salvation one must come to "know" in some special sense the one great and good God of Light who is *beyond* the universe, or cosmos of earth, planets, and stars. The demonic Archons, or astral deities, seek to impede the salvation of men, and man's duty is thus to *evade* rather than to accede to their influence. The following quotations from Mandean hymns and lamentations give some sense of the profound renunciation of, and antagonism toward, the astral and astrological doctrines that are a fundamental part of Gnosticism.

> The Seven [planets] and the Twelve [signs of the Zodiac] become my persecution . . .
> The Seven will not let me go my own path,
> How I must obey, how endure, how must I quiet my mind!
> How I must hear of the seven and twelve mysteries, how must I groan! . . .
> The evil ones [the stars and planets] conspire against me. . . They say to one another, In our own world [the physical universe] the call of Life shall not be heard, it shall be ours. . . . Day in, day out, I seek to escape them, as I stand alone in this world. . . . You see, O child, through how many bodies, how many ranks of demons, how many concatenations and revolutions of stars, we have to work our way in order to hasten to the one and only God.[47]

This negative revaluation of the astral deities is eloquently reflected in the Roman Servius' commentary on Virgil's *Aeneid*, in which as the soul moves down through the planetary spheres it accumulates evil characteristics from each planet—such as lust from Venus, avarice from Mercury, and wrath from Mars—and in Arnobius' report of Hermetic Gnosticim, in which "while we slide and hasten downwards to the human bodies, there attach themselves to us from the cosmic spheres, the causes by which we become ever worse."[49]

It is certain then that Gnosticism involved a self-conscious revolt against the perceived tyranny of the stars associated with earlier astral religion. It seems that this renunciation of astral fatalism and religion encouraged a much more profound, though indirect, renunciation of the underpinnings of science as well; but to understand how and why this should have happened we must briefly discuss the meanings of "knowledge" associated both with the tradition of astral religion and with Gnosticism.

In a curious way the astral determinism of Babylonian religion (and later Hellenistic astrology) is very similar to the materialistic determinism associated with much modern science. In both cases one seeks knowledge of observable, physical objects and events by relating those entities and events to other observable, physical objects. That the astral priests probably thought of the connection between the heavenly bodies and terrestrial events in a different way than modern scientists conceive the causal relationship between objects or events is irrelevant here. The important point of similarity is the common assumption, explicit for modern scientists but implicit for the Near Eastern astral priesthood and Hellenistic astrologers, that what really matters—both as cause and effect—is observable (phenomenal). Now, if one is unable to challenge the validity of this kind of knowledge, and is repelled by its consequences, there is no choice but to denounce the importance of phenomena and phenomenal knowledge and to claim the existence of some transcendent being or value, which must be known in some other way. This is precisely the position in which some ancient opponents of astral religion found themselves, though they were almost certainly not self-conscious about it in the way that a critic of modern science might be. They could not recognize the "incorrectness" of astral determinism and simply denounce it as wrong—something to be laughed at as an outgrowth, however sophisticated, of ignorant superstition—because it was a sophisticated product of the most powerful intellectual elite of their society. Thus, they turned to a belief in the existence of a transcendent, *alien* god beyond the cosmos governed by the astral deities. And they believed that enlightenment about *that* god must come from spiritual revelation and mystical participation rather than from observation of, or involvement in, the material world. While the body might be under the sway of the stars, the Gnostic believed that the immortal soul, which was vastly more important, could reach beyond the fates of the physical world. In this belief he was profoundly antiscientific, not self-consciously and by intent but antiscientific nonetheless. For he rejected the phenomenal world, and therefore phenomenal knowledge, as something unimportant at best or, perhaps, even inherently evil. The phenomenal world is (by definition or convention) the domain of science, and scientific knowledge is (by definition or convention) phenomenal knowledge; thus, there is an inherent conflict between Gnosticism and science as it has been defined in the Western world. Through a curious dialectical process, then, the scientific element of astral religion produced, or at least encouraged, its own antithesis in the Gnostic vision of knowledge.

It is extremely important here to emphasize that I am not in any way arguing that the emphasis on spirituality, ethics, and even salvation (which appears in Gnostic cults) derives from some kind of revolt against a "materialistic" astral religion. Nor do I want to assert that the notion of "revelation" developed in opposition to a notion of objective empiricism associated with astronomical science and astral religion. Salvation and ethics were central doctrinal concerns of earlier Persian religion with its astral features and of contemporary Judaism; and the very notion of revelation is central to the long Near Eastern tradition of omen interpretation and astrology in which truths are revealed through the agency of entrails or celestial objects. What is crucial for the Gnostics, and later for Christianity (which incorporated certain Gnostic sensibilities even though it condemned the Gnostic assertion of the existence of evil gods identified with the material world and separate from and in conflict with the one true God), is that opposition to astral religion forced a *separation* between the physical world and the spiritual one and a *distinction* between revelation connected with the senses and another kind of transcendent spiritual revelation that could come from the alien (i.e., outside the universe) God alone. The assumption of a fundamental opposition between this world and another, an assumption unknown to the Sumerians, Babylonians, Persians, or the early Stoa, signals a radical and lasting element of Western intellectual life.

Even this assumption, of course, may have had doctrinal origins outside of the tradition we have discussed so far. In particular, it is clear that many Gnostic formulations of the distinction between a more significant spiritual realm and an imperfect—often evil—material realm borrow substantial elements from Greek philosophers like Parmenides and Plato. But Platonism was probably a powerful force within Hellenistic religious movements, including Christianity, precisely because it responded systematically and directly to an inherent religious need associated with cosmology. It is almost certainly for this reason that Plato was known in Hellenistic culture and early Christendom almost exclusively through his single late cosmological treatise, the *Timaeus*, rather than through the numerous political and ethical writings that form the basis of his modern reputation, and were undoubtedly more central to Plato's own dominant concerns.

SUMMARY

Most of the attitudes and activities that characterize science emerged first in connection with the accounting and engineering problems of the earliest urban civilizations of the Near East. Moreover, they seem to have been combined into a more or less distinctive cultural specialization in the scribal schools of ancient Sumer and Babylon, where evidence suggests that at least some scholar-teachers sought systematic mathematical knowledge that went beyond the practical needs of most of their students.

Perhaps the most significant product of Mesopotamian scribal learning was a system of positional astronomy, which gradually increased in sophistication and predictive power from the early second millenium B.C. until nearly the beginning of the Christian era. The products of this mathematical-astronomical tradition were incorporated into Mesopotamian religion, producing major shifts in religious emphasis. It dissociated the religious calendar from its agrarian roots, strengthened the focus on omenology and fatalism, and generally produced an increasingly esoteric religion that lost its capacity to respond to the daily wants and needs of the population.

At one level the growing predictive ability of Babylonian astronomy, fused with the increasing astral focus of Babylonian tradition, produced an important and long-lived feature of Western intellectual life—horoscopic astrology—which was codified in Hellenistic Greece and played important roles in Roman culture, in Medieval Islam, and in Christian Europe during the Renaissance.

But on a second level, fear of (and opposition to) the esoteric science-based astral determinism of Babylonian religion encouraged the spread of religions with an extreme dualistic emphasis. The determinism of the physical world that was linked to astral determinism was associated with evil, and a distinct spiritual and transcendant world was posited and invested with positive moral attributes. Though this dualistic emphasis had probably been a feature of Persian religions before their interaction with Babylonian astral religion, it was rapidly adopted and incorporated in a whole group of Gnostic religions whose most pervasive feature was a repugnance toward astral determinism. In its Gnostic form the earlier confrontation between the good and evil powers became a confrontation between a good "spiritual" world and an evil "material" universe. As we shall see, these Gnostic attitudes played a major formative role in shaping early Christian theology.

Pre-Socratic Science, Theology, and Political Thought in Classical Athens 3

If we can properly speak of "scientism" as scientific attitudes and modes of thought extended and applied beyond the domain of natural phenomena to a wide range of cultural issues that involve human interactions and value structures, then we can say that Western culture was first infused with scientism in Hellenic Greece, even though it is clear that scientific knowledge had yet had little material impact upon society. There may have been periods of more intense and widespread scientism over the past two and one-half millennia, but it has seldom been more dramatic in its effects than between the early sixth and middle fourth centuries B.C. in Athens.

Ancient Near Eastern scientific activity had its greatest impact upon society through one of its products—the impressive predictive power of astronomical theory. Furthermore, although virtually every member of Babylonian society had his or her beliefs and activities directly affected by astronomical doctrine, the activities and modes of thought that produced Babylonian astronomy remained the provenance of a small specialist group. Even scholars who developed mathematical astronomy and astral religion did not extend their techniques to independent investigations of other subjects.

When we turn to Classical Greece the relations between science and society are vastly different. The early Greeks produced no science with substantial predictive ability. Nor was there in pre-Alexandrian Greece any small specialist group to whom scientific activities were confined, or who pursued predominantly scientific aims. Early Greek scientists tended to be men of affairs—politicians, merchants, admirals, generals, poets, and a few physicians—whose scientific activities were carried out largely in the absence of any clearly established institutional settings. Nothing even vaguely resembling the Mesopotamian educational systems existed to sustain learned activities in Greece until the fourth century B.C., and by that time scientific thought had had its greatest impact upon Greek society.[1] Incredibly, the amateur efforts of a handful of Greek natural philosophers, unsupported by strong specialized institutions, had a tremendous impact upon Hellenic culture, and through it, upon our own.

THE FOUNDATION OF HELLENIC RELIGION, POLITICS, AND MORALITY IN THE LATE BRONZE AGE

There is archeological evidence of continuous habitation at the site of Athens and other major Greek cities from as early as the third millenium, but the foundations of classical Greek culture were predominantly laid during the Mycenaean period. This lasted from about 1600 B.C., when a wave of literate, Greek-speaking invaders from the north arrived and established themselves as rulers over the indigenous population, until about 1200 B.C., when Mycenaean culture was destroyed—apparently by a new wave of preliterate invaders.

For nearly four hundred years after 1200 B.C. we have no written (and very limited archeological) records of Greek society. Then, when writing was again developed and a prosperous culture is attested, economic life and most social institutions had become substantially different from those of the Mycenaean age. There was, however, one all-important holdover. Superficially this holdover was both insignificant and uncharacteristic. It was a kind of distorting mirror of Mycenaean religious, political, and moral values embodied within two epic poems, the *Iliad* and the *Odyssey*; they were attributed to a single author, Homer, and probably written down from a variety of oral sources no earlier than the late eighth or early seventh century.[2] These two poems, however, along with the vastly less significant *Theogony* and *Works and Days* written by Hesiod during the late seventh or early sixth century, guided classical Greek religious, political, and moral life, both in terms of a general shared set of assumptions and values and in terms of ritual details.

Xenophanes, writing in the fifth century, was an opponent of Homeric values and inclined to blame Homer for anything he didn't like, but he was fundamentally correct in asserting of his own age

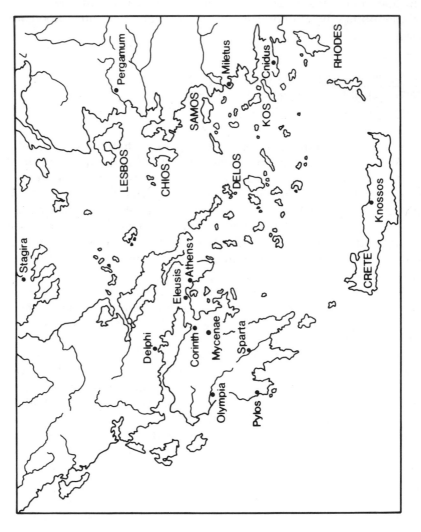

FIG. 3. Map of the Hellenic world.

that "all men's thoughts have been shaped by Homer from the beginning."[3] Homer's impact on classical political life was so great that a
twentieth-century analyst writes that "Homer not only provided the
foundation on which the polis and the people of the polis stood; he
was their companion, always present to them, always alive with
them, exerting a continuous and strong influence upon them."[4] Why
sixth- and fifth-century Greeks should have so revered the poetry of
an eighth- or seventh-century bard inculcating fourteenth- and
thirteenth-century values is a question beyond the scope of this work,
but they did—much as medieval Europeans revered the teachings of
Christ who had lived ten centuries before, and as twentieth-century
Africans revere the teachings of their seventh-century prophet,
Mohammed.

Because the Mycenaean culture as interpreted by Homer played
a dominant role in establishing the framework of the traditional
Greek values that early scientific thought confronted (and helped to
undermine and replace) we must consider what the Mycenaean culture was like and how its ideals were portrayed by Homer and the
traditional lyric poets who followed him up to the fifth century. Two
points should be kept in mind. First, it is only in retrospect and
against an awareness of economic and social changes, of which the
Greeks were only dimly aware before the rise of science, that we
easily distinguish between Mycenaean culture and that of Archaic
Greece. For Homer and his listeners, the Acheans (Mycenaean
Greeks) seemed to be their heroic but immediate ancestors; and the
values of the heroes were adopted—at least publicly and officially—
as their own. Second, in Homeric society, as in ancient Near Eastern
society, religious perceptions pervaded all aspects of life from the
interpretation of natural phenomena to the sanctioning of political
and social arrangements.[5] But within Greek society none of the
Eastern corporate specialties, such as priesthood or scribe, overlay
the common perceptions. Thus, an almost complete fusion of religious, political, and physical understandings was a central feature
carried into classical culture by the Homeric tradition. Lowes
Dickinson has expressed the situation very nicely.

> If there was no separate church, in our sense of the term, as an indepen-
> dent organism within the state, it was because the state, in one of its
> aspects, was itself a church, and derived its sanction, both as a whole
> and in its parts, from the same Gods who controlled the physical
> world.[6]

The reason for emphasizing this integration of religious concerns with politics and cosmology is to suggest why, when traditional beliefs about the origins and governance of the physical world were changed, religious beliefs were also necessarily undercut and transformed, and the whole fabric of social institutions lost its traditional justification and had to find some new set of sanctions.

What then was Homeric society like, and what were its dominant values? The ruling class of the society (the only class about which Homer cared) was a warrior aristocracy limited in each locality to a few families that played a leading part in the life of the city and its environs. Each family traced its descent to a single founder, usually a god or hero, and the only claim to membership in the ruling aristocracy was that of blood relationship. We can see the Homeric emphasis on an aristocracy of blood most clearly as it was being eroded during the archaic period. The aristocratic poet, Theognis, laments,

> Horses and asses and sheep we value according to breeding, Cynus, and wish them to be bred from the finest sires. Yet our nobles gladly accept bad women, of low birth, to be their wives, if only the dowry is great . . . the base now mate with the noble, the noble with the base; money confuses the breed, so do not wonder, Cynus, to see the breed of our Townsmen daily grow worse; for excellence is mingling with worthlessness.[7]

In each place one of the local aristocracy functioned as a king, directing the city in war and peace. Kingship, so far as we know, was not hereditary but was open to any member of the aristocracy who could garner enough power through personal prowess, skill in negotiation, and/or by marriage ties. Thus, the predominant relationships among members of the aristocracy were related to power. With the exception of certain codes of conduct concerning the treatment of one's social equals, the will of the more powerful was tantamount to law. This aristocratic social structure with its highly class-conscious code of behavior, its focus on power, and its association of the will of the mighty with law is reflected in Homeric religious thought.

Many centers of Mycenaean society were located in moderately rich agricultural regions, and the economy was basically a self-contained agrarian one. Some of the aristocracy were probably farmers themselves, and all were landowners; but it is almost certain that most agricultural work was done by others—by the indigenous conquered population, slaves, or retainers of the aristocracy. The rulers were predominantly warriors, and probably pirates as well. Thus, most of the human values of the society embodied in Homeric literature (especially the *Iliad*) were those of the military man— vigor, tenacity, physical power, aggressiveness, loyalty to com- patriots, and hardness verging on brutality.

Finally, although Mycenaean ruling class culture centered about the palace and used writing to keep track of commodities, it was vastly less complex and on a much smaller scale than the urban cultures of the Near East. There was no semiautonomous class of scribes and intellectuals. Little or no value was placed on reason and learning except in the immediate context of practical action. If any intellectual values were held at all, they were cleverness, shrewd- ness, and deviousness rather than extensive knowledge. Wisdom was seen as something definitely of the heart and will rather than of the head and intellect.

What was Homeric religion like, and how was it related to society? Homer's gods reflect many of the characteristics of the early natural deities of ancient Sumer. We see this nicely, for example, in the *Iliad*. Achilles is trying to light the funeral pyre of Patroclus, but the fire fails to catch for lack of a wind. So Achilles prays to the winds to come.

> Standing aside from the pyre, he prayed for the two winds of the North and the West and promised them fair offerings, and pouring large libations from a golden cup besought them to come that the corpse might blaze up speedily in the fire, and the wood make haste to be enkindled—and they came with a mighty sound, rolling the clouds before them. And swiftly they came blowing over the sea, and the wave rose beneath their shrill blast, and they came to deep soiled Troy, and fell upon the pile, and loudly roared the mighty fire.[8]

Here we are reminded of the character of nature gods—their ability to serve men and their need to be appealed to and bribed, or rewarded for their services.

Another passage—this time from the *Odyssey*—indicates a subtle but important difference between the Homeric Olympian gods and those of the ancient Near East. In this passage Poseidon, god of the sea and of earthquakes, sees Odysseus where he doesn't expect him to be.

> Now the God, the shaker of the earth, on his way from Ethiopia espied Odysseus afar off from the mountains of the Solymi: even thence he saw

him as he sailed over the deep; and he was yet more angered in spirit, and wagging his head, he communed with his own heart. Lo now, it must be that the Gods at the last have changed their purposes concerning Odysseus, while I was away among the Ethiopians. And now he is nigh to the Phoenician land, where it is so ordained that he escape the great issues of the woe which hath come upon him. But methinks that even yet I will drive him far enough in the path of suffering.

With that, he gathered the clouds and troubled the waters of the deep grasping his trident in his hands; and he roused all storms and all manner of winds. . .[9]

The willfulness of Poseidon here is able to subvert the decision of the council of gods—at least temporarily—just as the willfulness of a single member of the Homeric aristocracy might be able to subvert the intent of the collective leadership. Homer's gods reflect the ruling class in many ways; this is just one. No Near Eastern god could have flown in the face of collective decisions like this.

Both the Near Eastern and Homeric gods, of course, mirror the standards of morality of a semifeudal ruling class. Though the major gods do not poach upon one another's consorts, for example, their lusty attentions toward lower-class women—in this case mortals or minor deities—are seldom thwarted and never punished. Similarly, though the gods usually act honestly and honorably toward one another, even while fighting and bickering among themselves, they are not above using deceit when they deal with men or their distinct inferiors among gods.

The class consciousness of Homeric culture had one more major reflection in religion that was crucial in shaping traditional Greek attitudes: it was critical for a member of a feudal, hierarchical society to know his proper place and to stay in it, having thoughts proper to his station and behaving in an appropriate manner. Thus, in Homeric religion and morality the line between mortals and gods is very clearly drawn, and nothing is calculated to bring the wrath of the gods down upon a human more rapidly than overweening pride or *hubris*—the attempt to become or act beyond one's station. W. K. C. Guthrie makes the point with special clarity in talking of the relation of humans to Zeus, the king of the gods.

Before him, man stands helpless as a creature of a lower order altogether. He is immortal, man is mortal; he is all powerful, man is weak. He is a being entirely external to man, and to get into the right relationship with him it is necessary to proceed accordingly, acknowledging his supremacy and placating him with offerings and worship. He is simply a ruler who will brook no rivals. Even another God, should he set himself up against Zeus, could expect no better treatment than to be seized by the leg and hove out from Olympus, to suffer the nearest to death that an immortal god can suffer. How much worse would be the fate of a mortal who had thoughts above his station and in any way questioned Zeus' supremacy. His case is certainly not helped because from the moral point of view Zeus may stand no higher than the lowest of his human subjects. The two planes have nothing to do with morality, but are planes of power. It is in fact, a class distinction which separates man from his God, like that which separates the human King or Chieftain from the common people.[10]

There is another closely related and equally important religious implication of Homeric and archaic Greek class consciousness. Just as man stands clearly separated from the gods with no possibility of god-like immortality after death, man's station is clearly above that of other living beings; it would be totally inappropriate to interpret human behavior solely in terms of that proper to lower beings. Hesiod makes this clear in a particularly pointed discussion of a fable about the hawk (which symbolizes tyranny) and the nightingale (which represents the poet). A hawk seizes a nightingale in his talons; the victim complains, and the hawk responds by saying, "I am higher and mightier than you. I can drag you whithersoever I will, and can devour you if I will. He is a fool who tries to withstand the stronger; he suffers pain besides his shame."

After the fable Hesiod launches into a sermon on the blessedness of justice as a buffer against raw power, which he concludes by pointing out that "fishes and beasts, and wild fowl devour one another, for *right* is not in them; but to mankind Zeus gave right."[11]

The notion that man was fundamentally and radically distinguished from the beasts by the existence of human justice (or right) reflected in divinely sanctioned law, was a fundamental traditional belief—one that was severely challenged by scientific ways of thought.

CHANGED CIRCUMSTANCES IN THE ARCHAIC AGE

When Greek society emerged from the Dark Ages around 800 B.C., economic and social circumstances were greatly changed for a variety of reasons. Perhaps the most important reason had to do with modifications in military technology, which appeared just about the time of renewed literacy. The major military device of Mycenaean society had been the horse-drawn war chariot, but this was now replaced by the phalanx, a tightly formed group of foot soldiers (hoplites) as the basic military unit. As a consequence, the importance of a wealthy warrior aristocracy was drastically lessened, for now any man who could afford the relatively inexpensive equipage of the hoplites was—at least potentially—the military equal of the wealthiest member of the old ruling class. Perhaps equally important in cities like Athens, where the navy became critically important, even citizens who could not afford to be hoplites could row in the ships and be of military significance.

The Athenian navy also reflected the growing importance of another segment of society, the merchants. During the Bronze Age, trade had been accorded little significance by the aristocracy, and in archaic Greece the old attitudes were officially maintained. But trade was becoming increasingly important. By the end of the Dark Ages mainland Greece and the Peloponnesus were unable to produce enough food to support their increasing populations. Some relief came through colonization—to the east along the coast of Asia Minor and up into the Black Sea, and to the west in Sicily and southern Italy. But now trade in basic foodstuffs was added to an older trade, which had brought metals and other raw materials into Greece in return for things like fine pottery. Merchants and traders became increasingly essential to society and increasingly wealthy. Though largely excluded from extensive political roles and from higher places in the army because they were not from the traditional aristocracy, merchants supported much of the Athenian navy by purchasing, outfitting, and paying for the crews of ships. Thus, the state owed a largely unacknowledged economic debt, and an often recognized military debt, to a powerful and wealthy class that was essentially ignored in the Homeric tradition.

The old emphasis on a warrior aristocracy of blood was increasingly inappropriate to a society that depended for much of its existence on the talents and efforts of its lesser citizenry and merchant classes. This disparity between ideology and reality was reflected both in literature and (more importantly) in open political conflict between the ruling families and the more common citizens (the *Demos*) in numerous cities. In some places, as in Athens, political reforms gradually led to an increasing democratization of polis life, and the emigration of dissidents to form new colonies provided another safety valve. But elsewhere, as at Sparta, the old institutions were maintained with a consequent near total divorce between political and economic life. The military aristocracy claimed all political power while all economic functions were carried out by *helots*, a disenfranchised—but frequently prosperous—class, bound like later medieval serfs to their functions. Regardless of what happened to the forms of political institutions, almost everywhere, and especially at Athens, a more even-handed system of positive law began to replace the older unwritten conventions that had countenanced the exploitation of the lesser citizens by the aristocracy. In many places something approaching equality before the law for all male citizens came to exist.

Justice—especially protection of the weak from oppression by the strong—was a major social concern, and this led to a major crisis in Homeric religion. If it were no longer acceptable for the hereditary aristocracy to claim its feudal rights over the lesser citizenry in sexual matters, for example, then how could one condone the sexual license and blatant immorality of the Homeric gods? Even more important, though less obvious, how could one condone the gods' willful transgressions of communal decisions? Such behavior was tantamount to ignoring the laws and thwarting justice.

The official state religion remained focused upon Homeric gods; but from the early seventh century until the collapse of Hellenic culture there was widespread criticism among intellectuals, and there was a growing emphasis upon mystery cults that either closed the gap between the gods and human beings, or focused on the possibility of salvation through living a moral life. This movement was signaled by the incorporation of Dionysian cults—which promised a kind of temporary fusion of humans with gods—into civic religion; it continued with the appearance of the salvation-oriented Orphic doctrines during the sixth century, and eventually led to the rapid acceptance of Eastern astral cults during the Hellenistic period.

Poetry—especially that of Homer which portrayed the great heroes and the gods—was explicitly understood by the Greeks as the basic educational medium of their society, the source of basic values and of models to be emulated. Yet by the sixth century it was already clear that to embrace many of the values of the old mythology and the Homeric tradition would be to weaken rather than strengthen the basis of polis life. Even among the most conservative poets there was much unease about the old stories. In retelling one of the classical tales Pindar (ca. 522–438) comes to the point where he would have to describe a fratricide among the heroes and breaks off, "Here I stop; not every truth may with advantage show its face, and silence is often the best path that a man can tread."[12]

This, then, was the Greek culture into which scientific thought was introduced. It was one in which political institutions were so poorly integrated with economic and military ones that, as Aristotle wrote in his *Constitution of Athens*, "there was civil strife between the nobles and the people for a long time because the poor had no political rights."[13] And it was a culture in which the entrenched sources of religious belief and values were so poorly integrated with the social conditions that even some of the most conservative spokesmen for traditional values expressed unease, and purged the old stories of their most repugnant elements.

I have laid out these circumstances at length because when I speak later of the way in which scientific thought undermined the older tradition it must be clear that I am not claiming that scientific attitudes were intentionally, ruthlessly, and single-handedly used to uproot and destroy a vigorous and coherent world view that could have continued without extensive modifications. There would no doubt have been a gradual—perhaps even a fairly rapid—erosion of traditional Greek religion, politics, and morality even without the rise of science in the form of pre-Socratic natural philosophy. What I do claim is that the intrusion of scientific attitudes and ideas into a collapsing intellectual structure accelerated the downfall of traditional beliefs, and was decisive in shaping and forming the religious, ideological, and moral traditions that replaced those grounded in Homer. There would certainly have been change without the birth of science. But the particular intellectual changes that occurred—changes that have continued to inform Western culture to the present time—were very much a product of pre-Socratic scientific thought and its critics.

THE SCIENTIFIC CHARACTER OF PRE-SOCRATIC PHILOSOPHY

I can think of no better way to introduce early Greek science than by citing and commenting on an extensive passage from Hermann Frankel's *Early Greek Poetry and Philosophy*.

So far as we know, pure philosophy, divorced from all extraneous associations, came into existence suddenly and without visible cause. From a Greek frontier land where the blood of nations was mingled [Ionia on the coast of Asia Minor], where the cultures of West, East and South met and interpenetrated in conflict, at a time when exhaustion from within and despotism from without threatened to cripple the adventurous Greek spirit—there and then arose, as if by a miracle, the new and wholly Greek world of thought. These same Ionian colonies had formerly moulded Homeric religion (which also was in its way a peculiarly Greek construction); from Ionia an epic climate of religious feeling had spread over the whole Greek nation: now it was the Ionians again who first pushed aside this religious outlook to give free passage to their own thinking. In the generations that followed, it was still for a long time the colonial Greeks of East and West who carried on the philosophical movement. Mainland Greece being behind, since, unlike the colonizing peoples, it had not made a clean break with the past, and could not therefore so easily make up its mind to radical innovations.

Radical in all respects that new Philosophy certainly was, which now set itself up as a discipline in its own right. In the cosmos and in life it sought intelligible laws, not random chance or personal caprice: Hence it could hardly take personal Gods as its starting point, and could have no common ground with the national religion. It did not deal in particulars, but in universals; did not communicate by symbols, but by direct statement. Hence it disclaimed myth, which had supplied Hesiod's [cosmological]—notions with part of their substance and more than part of their form. As true and avowed philosophy it cared for nothing of the purely ephemeral in past or present. At the same time it never simply accepted anything important, but strove to see the reality beyond the appearance which things present to us. In this respect it diametrically opposed that tendency which we can trace in [Greek] poetry—the poets encouraged men to accept transitory experience as being of ultimate validity; they represented existence as having only one layer; in our mortal life we should experience, enjoy and suffer all reality to the very dregs. The philosophers saw life and the world as many layered, and during the temporary silence of poetry they pursued a vigorous and spontaneous metaphysic.[14]

I have chosen to use Frankel's introduction to pre-Socratic philosophy here not only because he raises a whole series of issues about early philosophy that we need to explore but also because Frankel, who is predominantly a classicist and literary critic, cannot be accused of having the same kind of axes to grind that I have. If he characterizes pre-Socratic philosophy in precisely those terms that I have already used to define scientific attitudes, it cannot be because he wishes to force it into my mold. And if he emphasizes radical rejection of Homeric religion and traditional myth as the most significant aspect of pre-Socratic philosophy, it cannot be because his principal concern is to consider the major cultural implications of the scientific enterprise; his concern is to understand the major sources of changes in Hellenic religious sensibilities.

Frankel does not talk about pre-Socratic *science*, but rather about pure *philosophy*; and we should consider for a bit whether we are justified in claiming for scientific attitudes and activities what Frankel, among others, attributes to a general philosophical movement.

Looking at Frankel's characterization of pre-Socratic philosophy we see that he emphasizes first that it sought "intelligible laws," not random chance or personal caprice. (In terms of our definition of science in chapter 1, this was a search for order.) Second, he points out its new explicit and universal claims. (We called it a search for universally valid knowledge.) Third, he insists that it "cared nothing for the purely ephemeral." (We said that science limits itself to a concern with the phenomenal.) Finally, he claims that it "never simply accepted anything important, but strove to see the reality behind the appearance." (We said that science is a "search for the underlying form" that lies beneath immediate sensory experience.)

Each of the major defining characteristics that we offered for scientific attitudes reappears as one of Frankel's characteristics of pre-Socratic philosophy, and there is no defining characteristic of pre-Socratic philosophy that is not included among those of science. In describing the earliest Greek scientist-philosophers or philosopher-scientists, there seems to be no way to distinguish between what is meant by philosophy and what is meant by science.

Almost all traditional treatments of pre-Socratic thought written since the mid-nineteenth century presume that philosophy is both temporally prior to and more inclusive than science, and that generalized intellectual activity invariably precedes specialized branches of study. Given these suppositions it makes more sense to talk about pre-Socratic philosophy, and argue that the portion of pre-Socratic *philosophy* that deals with the subject matter of such later specializations as meteorology, biology, or geology can be considered scientific only by an ahistorically-minded individual who insists upon doing so. From this point of view, pre-Socratic science is properly philosophy applied to certain limited ranges of subject matter, and is at best a small portion of a more general movement.

Though it would be essentially impossible (and probably foolish) to disregard this view entirely, its assumptions seem to be accepted uncritically from Comtean Positivism (which insists that religious and metaphysical stages of knowledge must precede positive scientific ones).[15] For our present purposes it will be interesting to entertain an alternative interpretation based on the assumption that in human institutions generalized kinds of activity may grow out of particular ones, rather than vice versa. According to this point of view, philosophy as we know it may have involved a generalization and extension of the domain of application of a set of activities that are more appropriately designated as scientific. If this interpretation is correct, then the whole Western philosophical enterprise is far from the least important cultural institution to owe its very origins to scientific attitudes and activities.

This point of view had some significant and astute supporters before Positivist ideology came to dominate the history of science and history of philosophy. Not only was it Aristotle's basic position, as we shall see, but it was also well developed by Adam Smith in the eighteenth century. As a young man Smith wrote a series of essays entitled, "The Principles which Lead and Direct Philosophical Enquiries: Illustrated by the History of Astronomy," and "The Principles which Lead and Direct Philosophical Enquiries: Illustrated by the History of the Ancient Logic and Metaphysics." In these essays Smith insisted that the astronomy-physics-logic-metaphysics sequence was the temporal sequence of origin, and that logic and metaphysics both arose to answer questions that were suggested by the early astronomers and physicists.[16]

At this point we must face a complex question that has no clear-cut answer. Suppose that in ancient Greece an enterprise associated with what we *now* call scientific attitudes, addressed to answering questions about a subject matter which *now* falls within the domain of the natural sciences, did key a wide range of cultural changes. We must still ask whether it is really legitimate in this instance to speak of the impact of science upon culture. Earlier we associated science not only with a set of attitudes and a range of subject matters but also with a set of activities, saying that we should be warranted in attributing an impact to science only when the adoption and spread of attitudes and ideas was demonstrably associated with the latter. Was this the case in connection with pre-Socratic philosophy?

If we could delineate an unambiguous list of activities to be labeled scientific, this question could be easily answered. But scientific activities vary significantly from time to time and place to place. They are genetically related to one another rather than constant over time; and just as it might be difficult to say when the ancestors of a present organic species became that species, it is difficult to say at just what point the historical antecedents of scientific activities might reasonably be called scientific.

When we look in detail at the extremely fragmentary evidence that we have about the activities of Ionian natural philosophers, it seems that we can reasonably talk about them as being engaged in science. This is especially true to the extent that we accept the suggestions in chapter 2 that ancient Near Eastern mathematics and astronomy were unambiguously scientific. Thales, for example, is reputed to have brought the study of geometry from Egypt into Greece, and the earliest proofs of several geometric propositions are associated with his name.[17] In addition, he is credited both by Herodotus and Eudemus with predicting an eclipse of the sun (probably that of May 28, 585 B.C.).[18] Both the geometry story and the eclipse prediction story are improbable in their detail, but there is

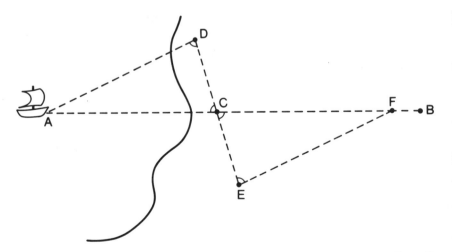

FIG. 4. Thales' method for discovering the distance of a ship at sea. From C sight along AC and extend AC to B. Construct line segment DE through C so angle ADC equals angle CEF; then AC will equal EF. Thales is reputed to have proved that two triangles are congruent when two angles and the included side of one are equal to two angles and the included side of the other. It is this proposition which guarantees the correctness of his method of determining distances at sea.

reason to believe that they reflect near truths; and if the stories have any grounds at all, they strongly suggest that Thales was acquainted with the Near Eastern astronomical and mathematical traditions and conversant with their technical details. Since we know from other sources that Thales was a merchant at Miletus, a major center of trade between Greece and the Near Eastern cultures, this acquaintance is at least plausible.

Similarly, Anaximander, the near contemporary of Thales at Miletus, introduced the first Near Eastern astronomical instrument, the gnomon (a vertical rod used to measure the altitude of the sun) to the Greeks; and he produced the first notable Greek map of the known universe—a map like those that had been produced since about 2,000 B.C. in Mesopotamia. Thus, he too is associated with the well-established Near Eastern traditions of scientific activity.

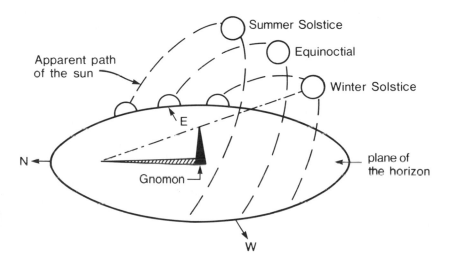

FIG. 5. Use of the Gnomon to determine equinoxes and solstices.

Neither Thales nor Anaximander (nor any other pre-Socratic philosopher) was engaged exclusively in scientific activities as some of their Babylonian contemporaries may have been, so it would be impossible to fully demonstrate that what we have called the scientific attitudes of the pre-Socratics were developed principally in connection with such activities. But those attitudes did appear initially in connection with astronomical, meteorological, geographical, and mathematical topics—precisely the topics associated with an ongoing scientific tradition—and the ancients themselves associated the rise of their new intellectual attitudes predominantly with astronomical and meteorological investigations and speculations. The few pre-Socratic philosophers who were not closely associated with mathematical and astronomical traditions were, like Empedocles, almost invariably physicians who were involved deeply with another set of unquestionably scientific activities—the empirical study of the character of human diseases.[19]

PRE-SOCRATIC SCIENCE AND ITS THEOLOGICAL IMPLICATIONS

Most discussions and classifications of the pre-Socratic philosopher-scientists have followed lines established by Aristotle, who was interested in how they accounted for the apparent flux of experience in terms of some unchanging principles. According to his classification scheme, the early Milesian philosophers, Thales, Anaximander, and Anaximenes, are best understood as monists; for each posited a single material substrate underlying the sensed variety of the cosmos. For Thales it was water; for Anaximander it was an indefinite amorphous substance (the aperon); and for Anaximenes it was air. Other pre-Socratics, like Anaxagoras, Empedocles, and the early atomist Democritus, are best understood as pluralists because they argued that there were two or more primordial material substances, and that nature demanded not only a material substrate but a cause of motion as well. Pythagoras and his followers form a special class because they believed that "the elements of numbers are the elements of everything, and that the whole cosmos is harmony and number."[20]

I shall return later to consider a few of the sweeping cosmogonic principles of the various pre-Socratics, but first I would like to focus on certain common characteristics of pre-Socratic thought. Many of the fragments attributed to or about most pre-Socratics are unrelated, or only marginally related, to grand cosmogonic speculations. Most of them offer explanations of particular natural phenomena, and many of the phenomena discussed deal with heavenly bodies, the weather, mathematics, and the nature of man—though there is an important class of fragments dealing with human society in post-Milesian times.

For the purpose of later reference, here is a short and characteristic selection of fragments dealing with a variety of topics:

[1] [on thunder, lightning, thunderbolts, whirlwinds and typhoons] Anaximander says that all these things occur as a result of wind: for whenever it is shut up in a thick cloud and then bursts out forcibly, through its fineness and lightness then the twisting makes the noise, while the rift against the blackness of the cloud makes the flash.[21]

[2] Anaximenes says that the earth, through being drenched and dried off, breaks asunder, and is shaken by the peaks that are thus broken off and fall in. Therefore, earthquakes happen in periods both of droughts and again of excessive rains; for in droughts, as has been said, it dries up and cracks, and being made overmoist by the waters it crumbles apart.[22]

[3] [According to Anaxagoras] . . .the sun, the moon, and all the stars are red hot stones which the rotation of the aether carries around with it. . . . The sun exceeds the Peloponnesus in size. The moon has not any light of its own but derives it from the sun. . . .Eclipses of the moon are due to its being screened by the earth . . . those of the sun to screening by the moon when it is new. . . . He held that the moon was made of earth, and had plains and ravines on it. . . .[23]

[4] [Democritus explained man's belief in divine power as follows] . . .the belief arose from alarming natural occurrences such as thunder, lightening, thunderbolts and eclipses, which men in their terror imagined to be caused by the Gods.[24]

The final passage of this group has a different character than the others, but it may help to explain why the others had an important bearing on religious questions. Traditionally, Zeus was god of the sky, responsible for such occurrences as thunder and lightning; Poseidon was the "earth-shaker" whose anger manifested itself in earthquakes and volcanic eruptions; and Helios, the sun, was the ubiquitous witness of all sacred vows. To account for such events as thunder, lightning, and earthquakes without calling upon the gods was, by itself, to challenge or erode a very substantial element of religious belief. But even more importantly, if one accepted the naturalistic explanations of such phenomena, one might easily be drawn into atheistic arguments like those of Democritus and lose faith in other aspects of divinity.

Such implications of the naturalists' ideas were recognized and deeply resented by substantial numbers of Athenians. Euripides pointed out that Anaxagoras had reduced "the all seeing Helios, who traversed the sky every day in his flashing chariot and was the awful witness of men's most sacred oaths, to the status of a lifeless lump of glowing stone,"[25] and this made such an impression that it was said to have caused an attempted expulsion of Anaxagoras from Athens. From the challenge to Anaxagoras until the trial of Socrates about fifty years later it was dangerous to be known in Athens as a man who speculated about the heavens; astronomical and meteorological thought were tied to impiety and atheism in the popular mind. Plutarch tells us that Athenians "did not tolerate the natural philosophers and chatters about things in the sky, as they called them, dissolving divinity into irrational causes, blind forces, and necessary properties. Protagoras was banished, Anaxagoras put under restraint and with difficulty saved by Pericles, and Socrates, though in fact he had no concern in such matters, lost his life through devotion to philosophy."[26]

Athenian intolerance showed up, as Plutarch suggests, not only in literary complaints but in the form of direct legal action. In about 440 B.C. a law that would impeach or exile "those who denied the Gods or taught about celestial phenomena" was introduced by Diopeithes. And in 411 B.C. Protagoras, who had been charged by the comic writer Eupolis of being an "imposter about the phenomena of the heavens," was brought to trial for impiety.[27] Though we have only Plutarch's word for the outcome of that trial we can be reasonably sure that his report was correct, for Protagoras died in a shipwreck on leaving Athens in about 410 B.C., and exile would have been the natural sentence had he been convicted. Moreover, we know that the Athenians ordered copies of his books to be collected and burned in the marketplace.[28] Far and away the most infamous persecution of ostensible impiety associated with "meddling in the affairs of the heavens" occurred in 399 B.C. when Socrates was tried and condemned to death.

In each of these cases (i.e., those of Anaxagoras, Protagoras, and Socrates) there were no doubt political reasons for the prosecutions, which were only marginally related to the avowed charges of impiety. But in each case an Athenian jury, which was probably responding at least in part to the charges rather than to the prosecution's hidden motives, felt that conviction was warranted because of the impious dangers of scientific speculation about the causes of heavenly phenomena.

Most pre-Socratic nature philosophers were neither atheistic nor agnostic, but their religious views tended to be very unorthodox. The two direct attributions that we have from Anaximander's cosmological writings demonstrate this clearly.

[Anaximander] says that it is neither water nor any other of the so-called elements, but some other *aperon* nature from which come into being all the heavens and the worlds in them. And the source of coming-to-be for existing things is that into which destruction, too, happens "according to necessity: for they pay penalty and retribution to each other for their injustice according to the assessment of time," as he describes it in these rather poetical terms.[29]

This seems to be the beginning of the other things and to surround all things and steer all, as all those say who do not postulate other causes, such as mind or love, above and beyond the infinite. And this is the divine; for it is immortal and indestructible, as Anaximander says.[30]

Anaximander's divine is quite unlike the traditional gods. It has no human form and cannot be appealed to for special favors. Its governance of the world is inexorable. In these ways Anaximander's divine aperon is abstract and remote from human experience. On the other hand, it has an important compensating virtue when compared with the Homeric gods because it is, above all, absolutely just—neither capricious nor immoral—and therefore is well suited to be a guiding principle for the new polis society in which equity in law and in commercial transactions was a crucially important issue.

Though Anaximander's religious ideas were more profound, and ultimately more influential, those of Anaximenes were of more immediate significance. Anaximenes and his near follower, Diogenes of Apollonia, asserted that air was the primary constituent of the universe, justifying their attitude in part by reference to the old notion of the breath soul. "Just as our souls, which were made of air, hold us together, so does breath and air encompass the world."[31] According to their cosmologies the various things that we sense are caused by the felting or compacting of air. But Diogenes carries this idea two steps further to argue that what we cannot sense (i.e., the soul) must be composed of the rarest and purest air, to which men give the name aether, and that this purest of all things is the divine.[32]

The notions of Anaximenes and Diogenes were adopted widely in connection with the so-called Orphic religions. W. C. K. Guthrie suggests that

the religious mind, supported by the edifice of theory, which the philosophers had erected on the foundations of ancestral belief, could, without difficulty, find comfort in some such train of thought as this. The air is of kindred substance to our souls, that is, it is alive. At its purest, it must be the highest form of our soul or life. Untrammeled by mortal bodies, it is clearly immortal, and the purity which it attains in the upper reaches . . . is such that it must be a very high and intelligent form of life indeed. What else can one call such life but God? The mystically minded could deduce from these same premises a satisfying promise of immortality for man. . . . It seems therefore that the air God of the Ionian Philosophers, with which we started, was not so far removed as one might suppose from the religious aspirations of men whose minds were very different.[33]

Whether or not we accept Guthrie's reconstruction of the Orphic mentality, there is substantial evidence of a semipopular association of both the soul and Zeus with the air or aether in the late fifth century, and following the natural philosophers' speculations. A funeral inscription dedicated to the Athenian soldiers who died fighting at Potidaea in 432 B.C., for example, reads, "Aether received their souls, their bodies, Earth."[34]

Moreover, Euripides associates Zeus with the aether in several passages. And Phelemon, writing in the early fourth century, finds the air-Zeus association a fit subject for parody: "I am he from whom none can hide in any act which he may do, or be about to do, or have done in the past, be he God or man. Air is my name, but one might also call me Zeus."[35]

Pre-Socratic science did, then, seem to authorize some religious beliefs. But these beliefs tended to be radically different from those of the Homeric tradition—beliefs in abstract, impersonal gods, or in natural gods so stripped of their old mystery and personality that they were subject to ridicule on the stage. During a period of terrifying rapid change and instability in such things as political institutions, moral standards, and economic circumstances, any challenge to old beliefs was likely to appear dangerous to the conservative element in society. And any person whose religious conceptions were at variance with those of tradition was likely to be charged with atheism and impiety. Thus, even where natural philosophy had a substantial theological bias it was often seen as being antireligious, and to seem antireligious was to seem dangerous.

XENOPHANES AND THE SOPHISTS ON THE ORIGIN AND NATURE OF THE GODS

One of the most illuminating places to look for the early impact of natural science upon Hellenic religious belief structures is in the brilliant poetry of the Ionian rhapsodist and theologian, Xenophanes. Born at Colophon not far from Miletus about 570 B.C., just as Thales was at the height of his powers, Xenophanes lived until he was over ninety; he ended his life around 475 B.C. at Elea in southern Italy, during Parmenides' lifetime. He earned his living as a professional singer of poetry—poetry that shows the clear stamp of pre-Socratic naturalistic speculation and self-consciously undermines traditional beliefs and values at every turn.

It was in Xenophanes' religious writings that the most dramatic consequences of the new thought appeared. Xenophanes insisted that there was but one God, "resembling man neither in mind nor in outward appearing. This single God, sees as a whole, hears as a whole, and thinks with his whole self."[36] Furthermore, God does not physically intervene in events, moving from place to place like the Homeric gods. "Always he remains in the same place, moving not at all: nor is it fitting for him to go to different places at different times, but without toil, he shakes all things by the thought of his mind."[37]

Such a radical departure from traditional thought had little or no immediate impact upon civic religious practices or popular cults among the common people. Yet Xenophanes' new conception of God was not ignored. If, for example, we look at Aeschylus' characterization of Zeus in the *Suppliants*, we can see it insinuated into the most powerful and conservative drama. According to Aeschylus, "he hurls mortals in destruction from their high towered expectations, but puts forth no force; everything of God's is without toil. Sitting, he nevertheless at once accomplishes his thought, somehow from his holy resting place."[38] Or, if we look later at Aristotle's concept of the unmoved mover we see a clear reflection of Xenophanes' God.

One might of course reasonably ask, what has all this to do with pre-Socratic science? Xenophanes was driven to his conception of God by a two-pronged attack upon traditional religion. The first attack was a moralistic one not necessarily associated with scientific attitudes or ideas; but the second and more crucial attack (because it differentiated him from numerous carping moralists) was an epistemic and anthropological one that depended on the naturalism and empirical element in pre-Socratic natural philosophy.

Like several other archaic poets, Xenophanes was appalled by the immorality of traditional values and traditional gods: "Homer and Hesiod have attributed to the Gods everything that is a shame and reproach among men, stealing and committing adultery, and deceiving each other."[39] But unlike most of the others, Xenophanes sought to understand and explain the origins of what was to him a blatantly unacceptable theology. The problem, as he saw it, was that men tended to conceive the gods in their own image: "Mortals consider that the gods are born, and that they have clothes and speech and bodies like their own."[40]

One might not consider this problematic, but experience with other cultures leads one to recognize that conceptions of the gods vary substantially from place to place: "The Ethiopians say that their Gods are snub-nosed and black, the Thracians that theirs have light blue eyes and red hair."[41]

The corrosive implications of this recognition can be seen in Xenophanes' ironic extension of the argument: "But if cattle and horses or lions had hands, or were able to draw with their hands and do the works that men can do, horses would draw the forms of the Gods like horses, and cattle like cattle."[42]

If the gods revealed things to us directly we could immediately recognize all of their absolute perfection. But, says the empiricist Xenophanes, the gods do not so reveal themselves and men must gradually attain knowledge and understanding as a result of their own very restricted experience. This is why men can initially attribute only the limited perfections associated with mortals to divinity. It is as if God had never given us the experience of honey, "then we should think that figs were very much sweeter."[43]

Xenophanes realized that our limited experience might be used to undercut his own projection of the divine nature as well as the conceptions of others. He admits that "no man has clearly seen it, nor will there ever be one who has eye-witness knowledge concerning the Gods and what I say on each and every matter." The gods are literally not sensible and thus not open to knowledge based on observation. Yet, Xenophanes argues, we can understand God through "acceptable inference." While only things that show themselves visible to mortal eyes may be unquestionably certain, we can have a probable knowledge of other things by keeping our assumptions analogous to known principles: "Let those things be assumed, which look like those that are certain."[44]

Xenophanes' argument is fundamentally the same as that of Bishop Joseph Butler in his eighteenth-century *Analogy of Religion*. We accept certain statements about natural phenomena as established through a purported combination of observation and rigorous inference. No such support is available for religious propositions, but any religious doctrine that has a close natural analog has, for that reason, presumptive evidence for its truth. And unless some strong argument can be made against it, it should be accepted and used as the grounds for action.

Werner Jaeger has very plausibly argued that Anaximander's aperon has virtually all of the characteristics of Xenophanes' God. It is divine, completely just, all embracing, unmoving, and so on. All that Xenophanes had to do was to marginally change the emphasis from material and cosmogonic to spiritual and moral to turn the aperon into his monotheistic God, or rather, to draw from Anaximander's substantial aperon support by analogy for belief in his God.

Whether or not Jaeger is justified in tying Xenophanes' God to the particular doctrines of Anaximander, it is clear that Xenophanes' theological arguments were deeply indebted to the naturalism and phenomenalism (or the empiricist element) in pre-Socratic science. What is also clear is that Xenophanes' self-consciousness about the cognative status of his religious beliefs, and his assertion of man's necessarily limited direct knowledge of the gods, could lead to serious doubts in one less committed to theism. The Sophist Protagoras, for example, followed Xenophanes' empiricist argument into a sceptical position respecting the gods.

> Concerning the Gods, I am not in a position to know of their existence or otherwise, nor their nature with regard to external manifestations; for the difficulties are many, which prevent this experience: not only the impossibility of having a sensory experience of the Gods, but also the brevity of human life.[45]

Even more destructive than Xenophanes' epistemic challenge to traditional theology was his analysis of the anthropomorphizing tendency of men. For this argument could easily be turned from one that merely challenged human imaginings of independently existing gods to one that claimed the gods were solely a human invention. In explaining their knowledge of the gods, some of the pre-Socratics and their interpreters thus explained away the existence of "real" gods entirely.

Such explanations took on three basic forms, each of which spawned a major anthropological perspective that lasted into modern times: (1) According to Democritus religious belief arose primarily out of fear of the unknown causes of the most violent manifestations of nature. (2) The Sophist, Prodicus, however, argued that men honored and began to worship things that nourish and benefit man—the sun, the moon, rivers, lakes, bread and wine, etc. Later they began to consider even the discoverers of new crops and useful inventions to be gods.[46] (3) The later Sophist, Critias, depicted religion as a self-conscious imposition by a wise ruler on his subjects to ensure their good behavior.

> I believe that a man of shrewd and subtle mind invented for men the fear of the Gods so that there might be something to frighten the wicked even if they acted, spoke, or thought in secret. From this motive he introduced the conception of divinity. There is, he said, a spirit enjoying endless life, hearing and seeing with his mind, exceedingly wise and all-observing, bearer of a divine nature. He will hear everything spoken among men and can see everything that is done. If you are silently plotting evil, it will not be hidden from the Gods, so clever are they. With this story, he presented the most seductive of teachings, concealing the truth with lying words. For a dwelling he gave them the place whose mention would most powerfully strike the hearts of men, whence, as he knew, fears come to mortals and help for their wretched lives; that is, the vault above, where he perceived the lightnings and the dread roars of thunder, and the starry face and form of heaven—fair wrought by the cunning craftsmanship of time; whence too the burning meteor makes its way, and the liquid rain descends on the earth . . . So, I think, first of all, did someone persuade men to believe that there exists a race of Gods.[47]

Critias' theory is remarkable because it somehow incorporates every recognizable implication of pre-Socratic nature-philosophy for religious belief. By tying the abode of the gods to the heavens and the realm of meteorological phenomena, Critias accounts for why naturalists' explanations of heavenly phenomena were seen to be dangerous by traditional believers. By emphasizing the deity's spiritual all-observing nature and its ability to see and hear with its mind alone, Critias responds to Xenophanes' critique of the old limited anthropomorphism of the Homeric gods. Moreover, Critias combines the fear interpretation of Democritus and the gratitude of Prodicus in explaining why the heavens are such a fitting place as the choice for the gods' abode. Above all, Critias' commentary shows how thoroughly destructive the thrust of pre-Socratic natural philosophy could be for traditional religious beliefs if it was pushed to its logical extreme.

NATURE VERSUS CONVENTION AS THE BASIS FOR STATE AND MORALITY

In the previous section we spoke of some of the Sophists, Protagoras, Prodicus, and Critias. But before we go on it is important to consider the basic aims of the Sophists as a group. In an important sense, the Sophists represent the initial specialization of education in Greece. Leon Robin, no friend to the Sophists, has nonetheless discussed the circumstances of their existence very effectively.

> The Sophistic movement of the fifth century represents a sum of independent attempts to satisfy the same needs by similar methods. The needs are those of a time and a country in which every citizen can have a share in the business of his city, and can obtain personal predominance by words alone; where the competition of individual activities gives rise to numerous conflicts before the popular law-courts; where every man wants to assert, in the eyes of all, the superiority of his . . . talents and abilities to rule his own life and that of others. Eagerness of thought and greed for enjoyment, the ambition to dominate and emancipate one's own activity, exuberant, superficial curiosity, *combined* with a supple strength and penetrating subtlety, enthusiasm, and versatility—these are the characteristics of the social environment, and above all, of the younger generation which aspires to make the best use of wealth and birth. It therefore needs masters; to teach it the art of individual success in social life, while sparing it the delays and disappointments of experience. This is the art, or principle, which the Sophists taught. They taught the science of "good counsel" in private and public affairs, that is, "virtue" in the precise sense defined above, and the means to "become superior" to one's rivals.
>
> Since the need which their teaching met was a need of all democratic cities, they went from town to town to the pupils who awaited them.[48]

Looked at from a slightly different perspective, in a society where aristocracy no longer brought automatic power, and where virtue, or *arete*, was no longer supposed to be inbred, the Sophists were hired by those aspiring to high standing in society to teach them techniques that could gain them power and success. The political assembly and the law courts were understood to be the centers of public life, whatever the economic realities might have been. So the ability to speak and argue effectively on matters of law and public policy was the principle prerequisite for gaining recognition. The Sophists' principal concerns thus lay in rhetorical techniques, and in what we would call the fundamentals of law—both civil and criminal—and public life.

Given the Sophists' dominant concerns, there was no obvious tie to scientific ideas or activities. In fact, the initial foci of pre-Socratic and Sophistic interest were far from one another. Yet in terms of both form and content the Sophists came to depend critically upon pre-Socratic thought. They consciously attempted to apply the methods of abstract thought that had been developed by the speculative philosophers for explaining the physical universe to practical questions of public and private life. And this led rapidly to a series of crucial questions about the origins and legitimacy of law and morality.

Natural philosophy had clearly undercut the traditional arguments for the divine source and sanction of both formal constitutional, or positive, law and of customary usage. The Sophists then became embroiled in a debate that has structured political theory ever since in the pre-Christian and in the post-Renaissance Western world. The central question about public and private life became the following: are human laws and social patterns grounded principally or exclusively in nature (i.e., *physis*) and therefore determined and understandable as part of, or in the same way as, natural phenomena; or are they dependent upon conventions and customs (*nomos*), more or less arbitrary, and freely chosen by men in order to maintain social stability?

Both kinds of answers drew most of their intellectual support from pre-Socratic theories. Both often found support from the same theory, just as radical socialism and conservative individualism claimed to find justification in the Darwinian theory in the nineteenth century. Both kinds of responses raised intense opposition from those who wished to preserve the city-state in anything like its traditional and religiously sanctioned form. And the conflict between the two led to increasing scepticism on the one hand, and increasing conservative antiintellectualism on the other.

On balance, the Sophists who argued for *physis* and against *nomos* as the primary sources of human laws, customs, and belief were both less numerous and less powerful original thinkers than their opponents. They fill a much smaller space in traditional histories of philosophy, but their impact—especially in generating enemies for the whole Sophistic movement—was out of proportion to their numbers and abilities. Virtually all supporters of *physis* were politically radical. Their arguments were not that extant positive laws were invariably grounded in nature, but that they should be; and that because written (or positive) law sometimes stood against a more fundamental natural law, it inevitably led to conflict that could be eliminated only by bringing written law into conformity with nature.

To us Antiphon seems perhaps the most interesting and appealing supporter of *physis*. He was a contemporary of Socrates and a general humanistic legal reformer with a hard streak of the cynical realist in him. This realist streak, moderated by a deep concern for justice, shows up in his most extended commentary on positive law.

> The most profitable means of manipulating justice is to respect the laws when witnesses are present but otherwise to follow the precepts of nature. Laws are artificial compacts, they lack the inevitability of natural growth. Hence to break the laws without detection does one no harm, whereas any attempt to violate the inborn dictates of nature is harmful irrespective of discovery by others, for the hurt is not merely, as with the lawbreaker, a matter of appearance or reputation, but of reality. Justice in the legal sense is for the most part at odds with nature. The Laws prescribe what we should see, hear, or do, where we should go, even what we should desire, but so far as conformity to nature is concerned, what they forbid is as good as what they enjoin.[49]

Antiphon explains a bit more about how present law and nature diverge by giving a series of examples that are particularly interesting because they represent real problems and not the kinds of straw men that so many such examples do. He confronts the issue of written laws that allow for self-defense but prohibit the defender from turning to the attack. And he considers the unwritten law that demands children to respect even parents who maltreat them. In both cases the extant law or custom is admirable in many ways, and Antiphon recognizes this. But, he argues, on balance such laws lead to greater pain than pleasure. They fail to prevent the original aggressor from pressing his attack and fail to avert the continued suffering of a mistreated child, and in practice they are usually formulated in such a way that they favor the oppressor more than the oppressed. Since the function of law and customs (according to Antiphon) should be to benefit humanity (by which he explicitly says he means to increase pleasure rather than pain), the reform of traditional law and custom is called for.

Though Antiphon's advice on the reform of law was sufficiently unorthodox to make it unpalatable to the conservatives, his naturalistic defense of social and racial equality must have been even more repugnant. There are no natural grounds, he argued, for distinguishing between those of noble and those of common birth or between those born Greek and those born barbarian, since by nature we are all made to be alike in all respects. "This can be seen from the needs which all men have. . . . we all breathe air with our mouths and nostrils, and eat with our hands."[50]

The way in which arguments for social equality were drawn from scientific themes is beautifully illustrated in a fragment from one of Euripides' last plays, the *Phoenissae*. The woman Jocasta pleads with her son to renounce ambition and to honor equality.

> What is equal is always a stable element in human life, but the less is always foe to the greater and ushers in the day of hatred. Equality it is who established numbers and weights for men and delimited number. Equal in the years' circuit are the path of dark night and of the sun's light, and neither grudges the other his victory. Shall day and night serve mortals and you not brook to give your brother equal share in the dynasty with yourself? Where in this is justice?[51]

Here equality among men is tied to the exact quantitative considerations that apply to calenderial astronomy, and human ethical considerations are seen as analogous to the course of inanimate phenomena.

Of course, not all of the apparent legal and ethical implications drawn from naturalistic thought look as acceptable to us as the arguments for social equality. In the *Laws* Plato summarizes an argument that he had attributed to Callicles in an earlier dialog, the *Gorgias*. According to Callicles, we learn from the animals that it is natural for the strong to dominate the weak. Among humans unnatural laws often mould men and teach that equality is just. But if some naturally strong leader should arise he would shake off such fetters and rise like a lion to show the glory of nature's justice. Thus, justice is associated once again with force and power. Plato complains,

> These views are held by men who in the eyes of the young appear wise—who say that the height of justice is a conquest won by force. Hence young men fall into irreligion, as if there were no Gods such as the Law enjoins us to believe in. Hence, too, outbreaks of civil discord as men are attracted to the right life according to nature, which plainly expressed, means a life of domination over one's fellows and refusal to serve others as Law and custom demand.[52]

Most Sophists, as we said, did not agree with Antiphon or Callicles and their ilk that legal and ethical principles are, or ought to be, drawn from the natural order. They argued strenuously that human affairs are not naturally determined and should not be. Instead, human institutions and values are guided primarily by convention (*nomos*). This does not mean, however, that the majority of Sophist ideas on politics, religion, or morality were traditional or conservative. Nor does it mean that their doctrines were not fundamentally shaped in response to the doctrines of pre-Socratic natural philosophy. The Sophistic supporters of *nomos* differed from traditionalists largely because they could no longer accept the notion that conventions were divinely inspired and sanctioned. They developed a new notion of the significance of convention in human affairs.

The framework for the Sophists' new discussions of the meaning of human conventions developed out of questions raised by pre-Socratic philosophers concerning the possibility of knowing or learning the nature of the universe. Science is a search for universally valid knowledge, and the notion of validity implies some criteria, either for testing the conformity of our knowledge to the reality it purports to illuminate or for otherwise judging the "truth" of our knowledge. The question of validity never arose—at least in a conscious and explicit way—in connection with traditional religious and mythopoetic understandings of the world. The appropriateness of a traditional myth was established by its very survival, and mutually contradictory mythopoetic accounts of any given phenomena or social practice seemed to be tolerated without generating overt unease. The notion of establishing some criterion for assessing the validity of knowledge, however, was a critical problem for the pre-Socratics. And from the outset it led in two closely related but separable directions.

One approach began by asking what character reality must have if any certain knowledge of it is to be possible. In such a direction lay the complicated and important doctrines of Parmenides and the Eleatics—doctrines that concluded that "being" or "reality" or "what is" must be undifferentiated and unchangeable, and hence totally inaccessible to sensory experience. Truth about what really exists can come only from logical analysis according to this tradition, and the flux of ordinary experience is subject to opinion alone, not to knowledge. Since human affairs are among things that are in flux, they cannot be determinant and must be subject to more or less arbitrary conventions, conventions that cannot be spoken of as *true* or *false*.

The other approach assumed that somehow sensory experience lay at the foundation of all knowledge at the same time that it recognized the fallibility of the senses. The central questions for this tradition were how to judge between apparently contradictory experiences and how to be certain that sense experiences are common rather than idiosyncratic. In such a direction lay the extensive discussions of Democritus and the atomists, as well as those of the major Sophist, Protagoras.

Democritus' central doctrine regarding this issue is best captured in the following fragment:

> By convention are sweet and bitter, hot and cold, by convention is colour; in truth are atoms and the void. . . . In reality we apprehend nothing exactly but only as it changes according to the condition of our body and of the things that impinge on or offer resistance to it.[53]

According to Democritus, our knowledge of the world is initiated by sensory experience even though reality consists only of the interactions of hard, solid little particles having no characteristics other than size, shape, and position. That the experiences of all men should be largely the same (i.e., universal) arises from the fact that we are similarly constituted and that we are all affected by the same real events. These events give rise to sensations when the atoms of external objects interact with the atoms that make up human beings, and we simply agree among ourselves to call certain kinds of sensations by certain names. But it is possible for a variety of reasons, such as illness or drunkenness, for different individuals to experience the same real events somewhat differently. Thus, although sensory evidence underlies our knowledge of reality, there is no strict one to one correlation between real events and our perceptions of them. The basic regularity of experience is overlaid by a contextually determined variability.

It was this Democritean emphasis on an irreducible element of subjectivity or variability in our experience that formed the grounds for some important Sophistic moral and political arguments.

The first to exploit atomist ideas for political and moral purposes was Archelaos, student of the natural philosopher Anaxagoras, teacher of Socrates, and the man traditionally held to mark the turning point of Greek philosophy from natural to human themes.[54] Archelaos argued that if hot and cold, bitter and sweet, have no existence in nature but are simply a matter of how we feel at any particular time, then we can hardly suppose that justice and injustice or right and wrong could have a more constant and less subjective existence. The principles governing the relationships between man and man are thus not absolute but are relative to the circumstances we find ourselves in, contrary to traditional belief.

To this argument Archelaos added an intense focus on the distinction between the role of chance in nature and that of design in human affairs. Again the argument was founded in atomist theory, which emphasized that the essentially chance or random initial motion of the atoms was the basic cause of all that occurs, but this time Archelaos claimed a basic disparity between natural and human events. The argument was most fully developed in the *Laws* (889a ff), where it was opposed by Plato. Moving essentially at random the elements somehow come together in appropriate ways generating the entire cosmos with its contents. Animals, plants, the seasons of the year, all owe their existence to these causes (i.e., chance operating on some original matter). Human art or design is a later, much less potent, and qualitatively different kind of force. Some early arts, like medicine and agriculture, merely assist the forces of nature, and may have substantial power. But political art and legislation are quite removed from nature. They are artificial, as are the gods, varying from place to place according to local customs. Because both the gods and the laws exist by convention and artifice, justice has nothing to do with nature but owes its existence entirely to design. And if justice is merely an artificial human creation, then it is subject to change at any time that humans choose to re-create it.[55] It is clear that such arguments could be used to justify attacks on tradition and thus may have encouraged political instability.

Protagoras offered an even more radical theory of knowledge than Democritus and Archelaos, but through an odd quirk its political implications were much more traditional. Even more than the atomists, Protagoras emphasized the primacy of sensory experience in our search for knowledge. According to Hermias he argued that "the standard and criterion of objective experience is Man, and the things that present themselves to the senses are objective appearances; those, however, that do not present themselves to the senses do not exist in any of the forms of being."[56] That is, Protagoras denied the atomists' claim of a more fundamental reality underlying the phenomena. Only sensory experience is real. But Protagoras well knew that phenomenal reality does not have the stability that characterizes the kinds of knowledge sought by the pre-Socratic naturalists; thus, he was forced to give up completely the notion that a constant nature of any kind underlay experience.

"Each of us," Protagoras argued, "is the master of that which has appearance and of that which has not, and one differs enormously from another, namely in this, that for one man some things have reality and appear to him, for another, other things."[57] It would seem that such a subjectivist doctrine must lead to complete anarchy, and this was the fear of some who sought to banish Protagoras from Athens and burn his works. But Protagoras sought to turn the doctrine into a force for moral and political moderation, if not conservatism. While there is nothing absolutely true or false for Protagoras, he insisted that there are better and worse conditions. This, he argued, is easy to recognize in connection with medicine. To a sick person many foods seem (and therefore are) bitter, but to a healthy person they seem, and are, opposite. Everyone agrees that the condition of health is better than that of sickness because it is beneficial to men. The physicians make men better by giving them physical health. By the same token, the Sophist seeks to make clear to men how to attain moral and political health.[58]

Physical health consists of living in harmony with one's surroundings so as to have sensations that "seem to have value," or we should say, "are pleasant." Similarly, moral and political health must consist of acting in a way that is useful and appropriate to promote harmonious and undisturbed life (i.e., one that is pleasant because it avoids the pain associated with such things as fear). Only when men agree to abide by laws can they be assured of their own safety and the fruits of their toil; only by living under laws can men raise themselves out of a state of savagery and into a society that offers a better life.

For Protagoras, then, the scientific emphasis on phenomena leads to epistemological relativism, and epistemological relativism in turn leads to the conclusion that there is no more important political and moral consideration than respect for law and custom, which have been devised to help men live in harmony. Men may recognize faulty laws (i.e., laws that do not promote social health) and seek to change them. But change must be sought only through proper legal processes, for to step beyond the framework of law is to promote the kind of chaos that corresponds to the greatest possible social illness. Protagoras thus ends in a position like that of Thomas Hobbes in the seventeenth century—he achieves political conservatism based on a total renunciation of the religious and epistemic belief structures that sustain traditional society.

Though competing arguments like those of Callicles, Archelaos, Antiphon, and Protagoras have played a central role in the modern Western tradition of political philosophy, the major immediate consequence of the Sophistic analyses of the foundations of law and morality was probably to engender a deep scepticism regarding any attempt to illuminate the subject. The generation of Athenians following that of Protagoras was a disillusioned one for many reasons. The whole structure of the city-state seemed to be crumbling away, pushed on by the defeat of Athens in the Peloponnesian War. But what made the situation even worse was that there was no stable intellectual world in which to find solace, or from which to reconstruct society.

Isocrates surveyed the intellectual scene and voiced the complaint of his whole generation against that of Protagoras when he pointed out that, with all their purported proofs, the only thing they really demonstrated was that "it is easy to truss up a false argument about whatever you like to put forward."[59] Thus an intense antiintellectual reaction grew up—a reaction that Socrates saw as a rejection of "logoi" (argument) of every kind.

ARISTOPHANES AND THE ANTISCIENTIFIC RESPONSE

In talking about the religious, political, and moral implications of pre-Socratic scientific ideas I have so far carefully avoided using two of the best, and certainly the most extensive, surviving sources of evidence. These sources, the *Clouds* and the *Frogs* of Aristophanes, are part of the antiintellectual reaction mentioned in the preceding section (though they are by an old conservative rather than by one of the disillusioned young), and are in some ways deeply unfair to the character of the arguments they attack. But they provide remarkably detailed parodies of arguments whose sources we can find treated only briefly in other more sympathetic places, and they attempt to portray the attitudes of ordinary citizens rather than philosophers. I have saved them in order to present them together and at some length, for their power would be lost if one could not see the unity and coherence of their perspective.

In Aristophanes we see the most direct depiction of the new, scientifically based learning as something that threatens to undermine Athenian religion and morality. Scholars have debated how seriously Aristophanes intended his audiences to take some of his ideas. The majority opinion, represented by such otherwise divergent scholars as Gilbert Murray and William Arrowsmith, is that Aristophanes did not have any real deep-seated bitterness toward Socrates, or toward the ideas he put in Socrates' mouth in *Clouds*. According to this view, Aristophanes was principally an early-day Don Rickles calling forth laughter by indiscriminately attacking and insulting friend and foe alike. Cedric Whitman, who acknowledges the serious moral tone of the *Clouds* as we know it, even suggests that in its original (now lost) version Aristophanes had treated the new learning favorably, and that is why the play was initially very poorly received. I tend to disagree with both attitudes. But my main argument is affected by neither Aristophanes' intentions nor by the possible existence of a quite different earlier version of the *Clouds*. However Aristophanes meant his words to be taken, and whatever might have appeared in the original version, it is clear that Aristophanes' fifth-century Athenian audience read out of the *Clouds* version that has come down to us a deadly serious questioning of the religious, moral, and political implications of the new learning.

The apparent conflict between the new learning and Greek traditions was of enough topical concern in 429 B.C. to make it an appropriate subject for treatment, presented in Aristophanes' *Clouds*, at the Dionysian festival. And the degree of this concern was so intense by 399 B.C. that Aristophanes' implied charges against his fictional Socrates in the *Clouds* apparently formed a significant support for the charges against Socrates: "You yourself have seen these very things in Aristophanes' comedy—a Socrates who is carried around in a basket and asserts that he walks upon the air, and a great many other absurdities, of which I am completely ignorant . . ."[60]

What was the specific content of the *Clouds* that could represent and inflame the unease of at least portions of the Athenian populace about the dangerous implications of the new learning? Here, for brevity, I must select passages from just one side of a portrayal that satirizes elements of both the new and the traditional in Athenian life and education. Very early in the play Strepsiades, the ignorant and slightly corrupt caricature of the traditional Athenian, enters the school of Socrates intending to learn the new methods of logic that will allow him to cheat his creditors. He meets Socrates, who has been making astronomical observations while suspended in a basket in the air, and he swears by the gods to pay any price for learning the new logic, to which Socrates replies, "By the gods? The gods, my dear simple fellow, are a mere expression coined by vulgar superstition. We frown upon such coinage here."[61]

Strepsiades then asks, "What *do you* swear by?" and Socrates offers to teach him the *real* truth about the gods, after which he introduces the chorus of clouds as the only gods there are. Strepsiades responds in two ways. First, he remarks that he thought that clouds were only fog, dew, and vapor. This is, of course, precisely what the Socrates of the play believes. His claim that the clouds are the only gods is not an attempt to dignify the clouds, but rather to naturalize and minimize the gods. Secondly, Strepsiades objects that Socrates has forgotten the great Zeus, to which Socrates responds, "Zeus, what Zeus? Nonsense, there is no Zeus." Again the straight man, Strepsiades, steps in to ask, if there is no Zeus, then who makes it rain? The ground has thus been laid for Socrates to bitterly satirize those naturalistic explanations of natural phenomena that had been offered by Milesian natural philosophers and were being taught by Sophists in fifth-century Athens.

Socrates begins by explaining that the rain comes from the clouds and that thunder occurs when the clouds bump into one another and roll over one another, moved not by Zeus, but by the wind.

Strepsiades professes not to understand, and Socrates tries another approach:

> Take yourself as an example. When you have heartily gorged on stew at the Panathena, you get a stomachache and suddenly your belly re-sounds with prolonged rumbling.

Strepsiades responds:

> "Yes, yes, by Apollo! I suffer, I get colic, then the stew sets to rumbling like thunder and finally bursts forth with a terrific noise . . . and when I take my crap, why, it's thunder indeed, *pa pa pax! pa pax! papapapa-pax*!!! just like the clouds. . . . And this must be why the names are so much alike: crap and clap."[62]

Following this magnificent parody, Stepsiades professes to be convinced, swears allegiance to the clouds, and vows, "If I met another god, I'd cut him dead, so help me. Here and now I swear off sacrifice and prayer forever."[63]

In this passage Aristophanes exposes the key issue in the play, the undermining of *nomos* (law, custom, tradition) by *physis* (nature). Socrates has robbed natural events of their dignity by somehow trivializing them, turning phenomena that had depended upon the will of Zeus into a grotesque manifestation of the farting of the clouds. Aristophanes seems truly to feel that such attitudes offend the gods and that Socrates and his ilk deserve to be punished for their offenses. To this end he finishes his play by having Strepsiades, finally cured of his venality, set fire to Socrates' school. As Socrates escapes Strepsiades beats him with a stick, yelling,

> Then why did you blaspheme the gods? What made you spy upon the moon in heaven?
> Thrash them, beat them, flog them for their crimes, but most of all because they dared outrage the gods of heaven.[64]

While the undermining of religious belief represents the most fundamental pernicious tendency of the new naturalistic learning, there are other important tendencies as well; Aristophanes focuses on the way a scientific approach, applied to human affairs, undermines morality and law as well.

Strepsiades gives up trying to learn himself and sends his son, Pheidippides, to Socrates. The son returns, gets into an argument with his father, and threatens to beat him. To this Strepsiades answers in horror that for the son to strike the father is contrary to all *tradition, law, and custom.* The son then offers to prove by argument that he has the right to strike his father, and the following passages ensue:

> LEADER OF THE CHORUS: Come, you, who know how to brandish and hurl the keen shafts of the new science, find a way to convince us, give your language an *appearance* of truth.
>
> PHEIDIPPIDES: How pleasant it is to know these clever new inventions and to be able to defy the established laws! . . . now that the master has altered and improved me and that I live in this world of subtle thought, of reasoning and meditation, I count on being able to prove satisfactorily my right to thrash my father. . . .
> Who made the Law? An ordinary man like you and me. A man

who lobbied for his bill until he persuaded the people to make it law. By the same token, then, what prevents me now from proposing new legislation granting sons the power to inflict corporal punishment on wayward fathers?. . .

If you're still unconvinced, look to Nature for a sanction. Observe the roosters, for instance, and what do you see? A society whose pecking order envisages a permanent state of open warfare between fathers and sons. And how do roosters differ from men, except for the *trifling* fact that human society is based upon law and rooster society isn't?[65]

Here we see again the battle between *nomos* and *physis* and how it is that the new rhetorical modes, which make the worse argument appear the better, are tied directly to the new naturalism.

In the past Hesiod focused on the difference between the fowls of the air and men. To men alone Zeus gave law. But now, against the older sense of divine and traditional sanctions for the law, we have Pheidippides focusing on the all-too-human politics of legislating and appealing to the demeaning parallels between the social order of chickens and that of men.

Strepsiades rails against this, telling his son: "Look, if you want to emulate the rooster, why don't you eat shit and sleep on a perch at night?" But his anger is to no avail, and he is defeated in this instance by the new learning.

I could multiply these examples several fold, for the *Clouds* presents the new learning as the source of decline in military vigor and physical culture as well as that of religion and fundamental morality; but I hope enough has been said to show how it is that Aristophanes' work could easily be read as an antiscientific broadside focusing on what Jean Jacques Rousseau later said of the scientists of the eighteenth century, that "these vain and futile declaimers go everywhere armed with their deadly paradoxes, undermining the foundation of faith, and annihilating virtue."[66]

So far, all of Aristophanes' attacks on scientism in religion and morality have closely paralleled arguments that had already been developed within the pre-Socratic and Sophistic traditions, though Aristophanes articulated them with special vigor and effectiveness. But now we follow Aristophanes' critiques in a slightly different direction.

At a supper celebrating Pheidippides' return from Socrates' school, the son refuses his father's request to recite from Aeschylus, declaring that the first great tragedian was "the most colossal, pre-

tentious, spouting bombastic bore in poetic history." Strepsiades
holds his temper and responds,

> All right, son, if that's how you feel, then sing me a passage from one of
> those highbrow modern plays you're so crazy about.
>   So he recited—you can guess—Euripides. One of those slimy
> tragedies where, so help me, there's a brother who screws his own
> sister.[67]

In the *Clouds*, Aristophanes does not further develop the contrast between Aeschylan and Euripidean tragedy; but some twenty-
four years later, in 405 B.C., he makes this the central focus of his last
major play, the *Frogs*. Here he emphasizes the uplifting, stirring, and
activating power of Aeschylan tragedy and compares it to what he
sees as a sense of disillusionment and paralysis brought on by
Euripides' realism, naturalism, and intellectualism—a realism,
naturalism, and intellectualism that is tied directly to Socrates'
scientizing influence.

The central plot of the *Frogs* is extremely minimal. After the
death of Euripides, the last great tragedian, Dionysus, the god in
whose honor the Athenian dramatic festivals were held, travels to the
underworld in order to bring back the poet who is characterized as
most "fruitful and generative." Initially he believes this to be
Euripides. In Hades he is convinced that he should judge a contest
between Euripides and Aeschylus to determine which author de-
serves this honor. In the process of the debate, which satirizes both
Aeschylan and Euripidean tragedy, Dionysus himself undergoes a
political education. By the conclusion, Dionysus decides that
because only the stirring works of Aeschylus have the power to rally
the populace of a demoralized Athens to action, it is he, rather than
Euripides, who should be returned to the city.

At the beginning of the contest Aristophanes recalls the conservative argument of the *Clouds*, which ties impiety to the new
scientific and naturalistic vision of the world. Aeschylus prays to
Demeter to give him strength for the contest. Euripides refuses to
approach the altar. When pressed, he displays the new "coinage" in
gods by tossing off the following lines:

> Ether, whereupon I batten, vocal cords,
> Reason, and nostrils swift to sneer,
> Grant that I may duly probe each word I hear.[68]

As the contest opens, Aristophanes also takes a casual swipe at the tendency of scientific intellectualism to seek quantitative and precise knowledge of everything.

> They'll bring straightedges out, and cubit rules, and folded cube frames, and mitre-squares, and wedges. . . . Line by line Euripides will test all tragedies.
> What, is it bricks they want?[69]

These bantering introductory shots, however, are merely the prelude to an attack upon scientific naturalism and reason as antithetical to the very aims of tragedy. At the core of this attack lie critical distinctions between ends and means, and between norms and descriptions. Aristophanes felt that though scientific analysis might tell what life *is* like, it could neither tell men what it *ought to be* like nor provide a sense of immediacy, urgency, and participation that lifts men from contemplation into action, like that derived from great music and poetry.

Euripides' first great claim is that he brought a realism and naturalism to the theater.

> I put things on the stage that come from daily life and business. Where men could catch me if I tripped; could listen without dizziness to things they knew, and judge my art.[70]

Aeschylus responds by pointing out that this very characteristic has made Euripides' plays publicly dangerous. In particular, Aeschylus attacks Euripides' treatment of women with its focus on what we would now call abnormal psychology, claiming that his treatment of Stheneboaea has encouraged suicide as a response to depression. Euripides defends himself with the scientific intellectuals' standard response, arguing that he is not responsible for inventing his characters—they are such as walk the street, or appear in traditional histories. Very true, responds Aeschylus,

> But the poet should hold such a truth enveloped in mystery, and not represent it or make it a play. It's his duty to teach, and you know it. As a child learns from all who may come his way, so the grown world learns from the poet. Only words of good counsel should flow from his voice.[71]

What men need is not an awareness of their own limitations, perversities, and corruptions, but rather a model to emulate—a lamp to follow rather than a mirror to contemplate. Thus Aeschylus proudly proclaims, "I taught you for glory to long, and against all odds stand fast."[72] It is precisely this kind of guidance that Euripides' scientific realism fails to provide.

Friedrich Nietzsche too saw a "slavish love of existence" which "left nothing great to strive for"[73] developing out of Euripides' works, and he called upon Aristophanes to support his claims.

> What Euripides takes credit for in the Aristophanean *Frogs*, namely that by his household remedies he freed tragic art from its pompous corpulancy, is apparent above all in his tragic heroes. The spectator now virtually saw and heard his double on the Euripidean stage. . . . It was henceforth no longer a secret, how—and with what saws—the commonplace could represent and express itself on the stage. Civic *mediocrity*, . . .was now suffered to speak, while heretofore the demi-god in tragedy and the drunken satyr, or demiman, in comedy had determined the nature of the language.[74]

The mention of language brings us back to the second of Euripides' major claims in the *Frogs*. For if he did not provide men goals to seek, Euripides was convinced that he provided an important instrumental education: "My figures didn't rave at random, or plunge in and make confusions," he says.

> I taught the town to talk. . . . I gave them cannons to apply and squares for marking verses: taught them to *think*, to see, to understand, to scheme for what they wanted. . . . This was the kind of lore I brought, to school my town in ways of *thought*, I mingled *reasoning* with my art and shrewdness, 'till I fired their heart to *brood*, to think things through and through. . . .[75]

Again, Aristophanes turns these proud claims back upon their author, attacking them on the grounds that *talk* and *intellectualized understanding* lead ultimately to a paralysis of will and failure to act effectively in the world. Aeschylus begins the attack by saying,

> True, you have trained in the speechmaking arts nigh every infant that crawls. Oh, this is the thing that such havoc has wrought in the wrestling school, narrowed the hips of the poor pale chattering children, and taught the crews of the pick of the ships to answer back to their officers' nose. How unlike my old sailor of yore with no thought in his head but to guzzle his booze and sing as he bent to the oar.

And Dionysus continues,

> But our new man just sails where it happens to blow, and argues, and
> rows no more.

The last word on this point, however, belongs to the final chorus
which ties Euripides, with his focus on speech and argumentation, to
both the renunciation of *true* music and poetry and to the Socrates of
the *Clouds*.

> Go, cast off music, poetry, And sit with Socrates and gas!—Leave the
> great art of tragedy And be—an Ass! Go, plunge in solemn argument,
> And spend a worthless afternoon in quibble, quiddity, and cant, And
> be—a goon![77]

Once more Nietzsche develops Aristophanes' insights. Euripides, he argues, demonstrates the fundamentally scientific attitude
that the value and beauty of a thing is tied directly to its intelligi-
bility—remember Poincare's statement that "intellectual beauty is
sufficient unto itself, and it is for its sake, rather than the future good
of humanity, that the scientist devotes himself to long and difficult
labors." In this connection Nietzsche follows Aristophanes in tying
together Euripides and Socrates—the *Clouds* and *Frogs*. It was
Socrates who first began to teach the world to value knowledge and
understanding above all else. Thus, Nietzsche writes, "Euripides
was, in a certain sense, only a mask. The deity which spoke through
him was neither Dionysus nor Apollo, but an altogether new-born
demon called Socrates. This is the new antithesis, the Dionysian and
the Socratic, and the art work of Greek tragedy was wrecked on it."[78]

Aristophanes' *Clouds* presented to us one of the first clear
portrayals of the conflict between the critical naturalism of science
and traditions of religion and morality, whose origins and justification lie in shared but frequently unanalyzed beliefs; his *Frogs*
presented to us the crucial struggle between the rational, intellectual, realistic man of science, whose primary goal is to *understand* or know what *is*, and the imaginative, visionary, involved poet,
whose primary goal is to *bring about* what *might be*. Aristophanes
clearly championed the latter over the former, not because he felt that
the man of science was intrinsically evil and worthless, but because
he felt the need for counterbalancing a growing domination of Athenian life by the scientific spirit of pre-Socratic philosophy as it was

applied to human affairs by the Sophists and by Socrates and Euripides. In this he shared (and first articulated) the basic value orientation of those who still speak today on behalf of the imaginative reason against the *domination* of modern life by the spirit of science; theirs too is a crying out to maintain some concern for poetic visions of the *ought* and ultimate values in a world dominated by empirical and rationalistic knowledge of the *is* and instrumental values.

SUMMARY

The basic argument of this chapter is that the scientific attempts of pre-Socratic thinkers to understand and explain natural phenomena were introduced into a world of economic, religious, political, and moral instability. Traditional modes of thought and sets of attitudes and assumptions were under attack, so it was possible for the newly emerging scientific attitudes and modes of argumentation to become key factors in establishing the direction of change within virtually all domains of intellectual life. The extension of scientific ways of thinking about nature (*physis*) into theological, ethical, and political issues was particularly easy because the lack of specialization among Greek citizens placed cosmological and meteorological speculation, as well as political activity and religious thought, within the proper domain of anyone ambitious enough to claim an interest. When thinkers turned their attention from physiology, meteorology, or astronomy to the origins of language and law, or to the significance of social institutions and religious beliefs, they naturally appealed to techniques of investigation and argumentation and to patterns of relationships that they had used or discovered in their studies of natural phenomena.

Initially, religious belief structures were most directly affected by scientific attitudes and ideas. As cosmogonic myths were replaced by naturalistic explanations to account for a variety of natural phenomena, the traditional gods were deprived of limited but very important functions. This challenge to isolated aspects of the gods' significance acted as a focus for intense and explicit criticisms of long-standing religious beliefs, which were becoming inappropriate to Greek culture for other, less easily expressed reasons. Naturalistic natural philosophy also led to a new self-consciousness about the status and sources of knowledge, and the resulting critical rationality and skepticism undermined traditional claims regarding divine roles, not only in guiding natural phenomena but also as sources of law and morality.

To replace older divine sanctions for politics and ethics, scientistic thinkers offered two basic kinds of explanation of social and moral foundations. On the one hand, some thinkers argued that one must seek the origins and justifications of all law and morality in nature (*physis*), patterning their analyses of human institutions upon those of natural phenomena. On the other hand, an even greater number of fifth-century intellectuals became disillusioned about the claims of natural philosophy because there were so many mutually contradictory systems, and because each scheme seemed to offer effective criticisms of all others. It might seem odd to speak of those driven toward scepticism by disillusionment as embracing scientism, but it is legitimate and important to do so because they were awed by the implications of natural philosophy and adopted a variety of scientific attitudes in spite of themselves. According to most members of this group, nature philosophers had shown that law and morality have no divine sanction. In addition, through their mutual criticisms, the nature philosophers had established (indirectly by their failure to discover it) that there was no knowable and invariable nature from which law and morality could derive their legitimacy either. With neither divine nor natural foundation, law and morality must simply be due to convention (*nomos*)—an arbitrary agreement among men designed to do no more than perpetuate society. Rights are nothing but what the powerful have over the weak, and laws are only to be obeyed when somebody is looking.

The real crisis of Athenian intellectual life brought on by these considerations was coincident with, and related to, an immediate political crisis—the decline of the Greek city-state, or polis, which was being accelerated by intense inter-city warfare. While it is virtually impossible to untangle the complex relationships between intellectual, political, and military problems, it is fairly clear that many conservative Athenians blamed the decline of their political and military fortunes on the scientistic intellectuals' undermining of traditional religion, law, and morality, and on the corrosive cynicism that they spawned. At the hands of the comedic poet, Aristophanes, an overt attack upon the new learning in the name of traditional values was effectively launched.

The Science and Scientism of Plato and Aristotle: Keystones of Western Intellectual Life

4

Like the aging Aristophanes, the young Plato found political and moral conditions in Athens at the end of the fifth century appalling; like Aristophanes he detested the growing impiety and scepticism connected with Sophistic grapplings with early scientific thought; and like Aristophanes he lamented the passing of old customs and institutions. But unlike Aristophanes, the young student of Socrates could not reject the new intellectual style in toto, nor seek a return to the great poetic tradition from Homer and Hesiod through Aeschylus and Sophocles. In fact, Plato launched a vicious attack against all poetry (and in a sense, against all tradition) in the name of knowledge achieved through philosophical methods derived from one of the early scientific schools—that of the Pythagoreans.

Most modern scientists and some historians would argue that far from being grounded in science, Platonism is fundamentally antiscientific. Such interpreters, like the empirically oriented physicians and atomists of ancient Greece, or like Xenophanes and Protagoras, claim that empiricism (i.e., direct sensory observation and/or experimentation) is *the* central characteristic of scientific activity. Science is and must be, according to them, predicated on the primacy of sensory experience. Plato bitterly opposed the empiricist assumptions of men like Protagoras, mainly because they seemed to lead inevitably toward sceptical conclusions. He argued that sensory knowledge was imperfect and untrustworthy, and that the world of sensory experience is but an imperfect and mutable copy of a *real*, perfect, immutable world, which is ultimately the only knowable one. From this point of view Plato blocked real scientific knowledge because he denied the primacy of phenomena. Consequently, it is very difficult for those who interpret Plato in this way to acknowledge that his perspective should in any significant way be called scientific, or be said to derive from science.

It is necessary briefly to challenge this superficial view of science and of Plato in order to gain a fair hearing for what follows. First, it is not true that Plato was uninterested in the ordinary phenomenal world in which we live and act. The fundamental motive for his development of an elaborate metaphysical theory was his desire to establish a just government in this ordinary world. According to Plato, although our reasoned knowledge can be only about the immutable and atemporal world of ideas or forms, that knowledge can be used to guide ordinary behavior because the ordinary world is in some sense a copy, or image, of the world of ideas. The relationships that hold in the world of ideas are, therefore, mirrored (or imaged) in the sensory world, even if only imperfectly. They can not be discovered in the sensory world because no absolutely invariable relations hold in that world; but once known, the forms or ideas can serve as a useful and valuable guide to true opinion about the sensory world.

The similarity of function served by Plato's ideas and by modern scientists' idealized physical systems can be understood by considering the way that the so-called ideal gas laws (which can be used to predict a wide range of phenomena) are in fact derived from the properties of a system of perfectly elastic spheres, that can be *realized* only conceptually, never in nature. Modern scientists, of course, do not ordinarily attribute a special level of reality or "being" to their hypothetical or idealized "physical" systems; and few, if any, would deny that they can have direct and well-founded knowledge about phenomena. So, for modern scientists the phenomenal world is not subject only to opinion. Thus the metaphysical assumptions of Platonism are not those of most modern scientists. Nonetheless, there are extremely important ties between scientific attitudes and activities and Plato's philosophical system.

Because Plato's metaphysical speculations—his so-called theory of ideas or forms and his dialectical method—were derived from early science, and because the entire Western philosophical tradition owes a tremendous debt to Plato's metaphysics, the origins of his methodological attitudes and his theory of forms deserve some attention here.

Plato was born into the Athenian aristocracy about 428 B.C., and was a cousin to Critias, one of the most able of the oligarchs established as rulers in Athens in 404 B.C. after Athens' final defeat in the

Peloponnesian War. As a young man he fully intended to enter politics, and become a student of Socrates, whom, he greatly admired. Because he had relatives in the oligarchy, Plato was invited to take part in the ruling, but soon the oligarchs unsuccessfully attempted to get Socrates to violate his moral principles in their support. When the brief reign of the oligarchs was over the new ruling group condemned and executed Socrates because of his ostensible impiety, and because of his very real significance as a political critic. Plato's hopes that the new rulers would bring a just government to Athens were dashed by the way they treated Socrates. His resulting disillusionment shaped the rest of his life and much of Western intellectual history. As Plato (or some very skilled and knowledgeable forger) wrote,

> When I considered this [the circumstances of Socrates' death], and the men who were active in politics, and made a closer study as I grew older of law and custom, the harder it seemed to me to administer the state rightly. For one thing it was impossible to act without friends and trustworthy associates, and these it was not easy to find at hand now that the city was no longer administered by the laws and institutions of our fathers. . . . At the same time the whole fabric of law and custom was going from bad to worse at an amazing rate . . . in the end I saw clearly that in the case of all existing states their government is without exception bad. . . . I was forced to assert, in praise of genuine philosophy, that only from that standpoint was it possible to get a true view of public and private right, and that accordingly the human race would never have respite from its troubles until either the true and genuine philosophers gain political control or else those who are already governing the states become, by some divine dispensation, real philosophers.
>
> It was with this conviction that I first went to Italy and Sicily. . . .[1]

Prior to Plato's departure for Italy and Sicily he had almost certainly written a series of dialogues—including the *Crito* and *Apology*—honoring his teacher Socrates and based largely on Socratic teaching and methods with little Platonic overlay. These early dialogues, for example, showed almost no interest in mathematics or natural philosophy; they invariably ended in an inability to resolve the questions posed (i.e., in a kind of skeptical paralysis), and showed no clear evidence of what was to become Plato's central metaphysical doctrine—the doctrine of ideas.[2]

Plato left Athens for his first extended travels sometime between his twenty-eighth and fortieth year. Sources differ on the year, but all accounts agree on at least two things—that he went to Italy, and that one of his aims was to visit certain Pythagorean philosophers. His visit with the Pythagoreans had a substantial impact on his work, as evidenced by a major change in the form and content of his dialogues, beginning with the *Meno* written during, or shortly after, Plato's first Italian visit.

PLATO'S SCIENTIZING RESPONSE TO CRISIS: THE THEORY OF FORMS

In order to understand the Pythagorean impact on Plato we must briefly summarize some of the major tenets of Pythagoreanism. Unlike most pre-Socratics, Pythagoras was the head of a religious cult or sect, and among the most critical religious doctrines of the sect was that of the transmigration of souls (metempsychosis). Like Zoroastrians and other eastern religious cultists, the Pythagoreans believed that the soul is not only immortal (i.e., that it does not die with the body) but that it moves from one body to another in the process of death and birth. If one lived a particularly pious life, following the proscription of meat in one's diet and studying philosophy, then one's soul would return in a higher bodily form—presumably with a higher rank in society. Ultimately, after a series of successful lives of increasing study and purification, one's soul might achieve perpetual union with the gods and escape from the cycle of repeated life and death.

Somehow associated with these religious beliefs was a cosmology that placed an extreme emphasis upon mathematics. In the most extensive ancient statement of the principles of the Pythagorean sect, Aristotle reports that

> the so-called Pythagoreans who were the first to take up mathematics, not only advanced this study, but also having been brought up in it they thought its principles, numbers, are by nature the first, and in numbers they seemed to see many resemblances to the things that exist and come into being. . . , since, again, they saw that the modifications and the ratios of the musical scales were expressible in numbers—since, then, all other things seemed in their whole nature to be modeled on numbers, and numbers seemed to be the first things in the whole of nature, they supposed the elements of numbers to be the elements of all things, and the whole heaven to be a musical scale and a number. And all the properties of numbers and scales which they could show to

agree with the attributes and parts and the whole arrangement of the heavens, they collected and fitted into their scheme; and if there was a gap anywhere, they readily made additions so as to make their whole theory coherent.[3]

When Plato wrote the *Meno* he began with a traditional Socratic question: Is virtue teachable? In the first section of the dialogue he went through the usual Socratic technique of reducing Meno to a state of total scepticism regarding even the meaning of virtue. But then Plato did something that differed radically from the pattern of all of his earlier dialogues. He began to offer a positive method for discovering knowledge. Beginning with the Pythagorean theory of the transmigration of souls, he argued

> since the soul is immortal and often born, having seen what is on earth and in the house of Hades, and everything, there is nothing it has not already learned. No wonder, then, that in the case of virtue and other things, it is possible for the soul to recollect what it once knew. For all nature is akin, and since the soul has already learned everything, there is nothing to prevent a man who has recollected only one single thing— *that is what we call learning*—from rediscovering everything for himself, provided only that he has the fortitude and doesn't weary of the search. So the searching and learning is only a matter of recollection.[4]

If we temporarily ignore the problem of how the soul could ever have come to know things in the first place, we can recognize that this theory of learning through remembering could be tremendously important to Plato because it provided a new hope that the sceptical methods of the older natural philosophers and Sophists, as well as of the early Socratic dialogues, could be overcome. When we realize that learning is just remembering what we already know we won't be "led astray by contentious arguments,"[5] but will have the courage to persist. Even more importantly, if learning is only remembering, it will make little difference how the memory is jogged, and a mere hypothesis, or speculation, might be adequate to bring our knowledge back.

Meno, of course, doubts the validity of this theory of learning and asks Socrates (Plato's mouthpiece here) for some kind of conclusive proof. Socrates challenges Meno to bring forward any of his servants, and says he can make the servant recall how to construct a square whose area is twice that of any given square. Then Socrates proceeds to ask the slave a series of questions, never telling him anything, always forcing him to make judgments. After something of a struggle, the slave recognizes that a double-sized square can be constructed by using the diagonal of the original square as the side of the new one. He has discovered this truth without any direct information from Socrates, so he must have known it previously and merely recalled it. Stated in slightly more modern terms, the soul has certain *innate* ideas that cannot be learned in a traditional sense but can be recalled with sufficient effort. This whole theory, of course, rests on the identification of all knowledge with what might be a peculiarity of mathematical knowledge.

J.E. Raven has nicely explained how the special character of mathematics suggested the whole theory of ideas to Plato.

> If, in a world in which we have never seen anything that is perfectly square or perfectly circular, we can nevertheless clearly visualize and exactly define the perfect square or the perfect circle, and if we can on this foundation build up a whole system of eternally true propositions, why should we not be able to do exactly the same in the field of ethics? Why should not perfect piety be on just the same footing as the perfect square? Why should we not be able to define perfect piety as precisely and formulate eternally true propositions about it too? . . . While the particulars of the world of sense are, as Heraclitus had said long ago, undergoing incessant change, these entities are eternally changeless; and so, while the particulars of the sense world can be objects only of

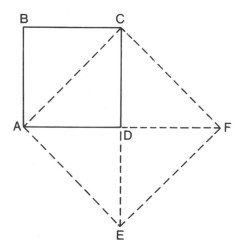

FIG. 6. Meno's problem: to construct a square with double the area of a given square. Area ACFE is twice area ABCD since it is composed of four triangles equal to triangle ABC, while ABCD is composed of only two such triangles.

opinion, right or wrong, this other class of entities are the objects of knowledge.[6]

After the discussion of learning as remembrance, Meno and Socrates return to the problem of discovering the nature of virtue and whether it is teachable in this new sense of teachable. Once more, through the character of Socrates, Plato argues that it is the method of hypothesis used by mathematicians that must lead us to the truth.[7] Throughout the great middle dialogues of Plato, the *Phaedo* and the *Republic*, the form and methods of mathematical thought continue to provide the basic imagery that supports both the theory of ideas and Plato's ideas about the nature of learning and education.[8] In the *Republic* Plato insists that years of mathematical training are necessary preparation for dealing with the most crucial political and ethical questions. Even toward the end of his life Plato continued to base many of his discussions upon mathematical concerns. This claim is supported by a famous report by Aristoxenus about one of Plato's last public lectures, a lecture on *The Good*.

> Everyone went there expecting that he would be put in the way of one or another of the things accounted good in human life, such as riches or health, or strength, or any extraordinary gift of fortune—but when they found that Plato's arguments were of mathematics and numbers, and geometry and astronomy, and that in the end he declared the *One* to be the *Good*, they were altogether taken by surprise.[9]

As soon as Plato began to emphasize the importance of hypothetico—deductive reasoning—he recognized that the initial hypotheses in any argument could be challenged. Most low-level hypotheses could be justified by arguing from some more fundamental hypothesis, but at some point, Plato acknowledged, one would have to stop arguing from hypotheses and provide a nonhypothetical justification of some set of first principles. The method for doing so Plato termed the "dialectic." If mathematical ideas and methods provided the principal stimulus to Plato's theory of forms and his early methods of argumentation, the methods of taxonomic classification from what we call the life sciences informed his later notions of dialectic.

Evidence that Plato placed substantial emphasis on problems of classification comes from at least two sources outside the Platonic dialogues themselves. One of the comedies of Epicrates, for example, portrayed the members of Plato's school, the Academy, as deeply involved in classifying trees, animals, and lettuces, and attempting to define the genus of the pumpkin.[10] And Aristotle quotes from a now lost work called *Plato's Classifications* in connection with a theory of the elements constituting the world. But the most important statements about his changing attitudes or emphases regarding philosophic method come from the later dialogues, especially the *Phaedrus* and the *Sophist*. In the *Phaedrus*, for example, Plato develops the method of collection and division that looks like the biologists' technique of collecting organisms into a genus on the basis of their similarities and dividing them into species according to their differences.[11] Then he displays his enthusiasm for the method by writing,

> I am myself, Phaedrus, in love with these division and collections, in order that I may be capable of thinking and talking; and if I reckon anybody else able to discern the one and the many as they are in nature, I follow him "as in the steps of a God." And moreover, those who have this ability I have hitherto called, God knows whether rightly or wrongly, by the name of dialecticians.[12]

Thus, throughout his career and in developing his most characteristic theories Plato drew principally upon concepts and methods that had been developed in connection with scientific attitudes and activities.

PLATO'S COSMOLOGY: THE *TIMAEUS*

Plato's dominant concerns were with politics and morality rather than with the subject matter of the natural sciences. Moreover, his fundamental assumptions downplayed the significance of sensory experience (i.e., phenomena). Yet Plato did write one late dialogue that dealt centrally with almost all aspects of the natural world. It is probable that this dialogue, the *Timaeus*, was intended principally to provide a background for two subsequent dialogues on Plato's favorite political and ethical topics.[13] But whatever Plato's intent, the later dialogues were not written as projected, and the *Timaeus* developed a life of its own.

In Plato's own words, the *Timaeus* was—and could have been—no more than a "likely story"[14] regarding the creation and structure of the physical universe. It is, however, a *tour de force* of current physical theory—a synthesis of elements from all major pre-Socratic schools with a new directive twist that makes it unique in its scope and explanatory power. A. N. Whitehead insisted that the *Timaeus* and Newton's *Principia* stand alone as the two greatest Western cosmological treatises, and such scientists as Werner Heisenberg have confirmed that judgment.[15] Whatever its intrinsic merits, however, the *Timaeus* must receive our attention because between about A.D. 100 and 1100 it was the only Platonic work available in the West, because it was taken as *the* central work of Platonic and Neoplatonic philosophy, and because it became a cornerstone of early Christian theology (see chapter 5).

We have already seen that Plato opposed the impiety associated with such pre-Socratic schools as those of the atomists and of Anaxagoras. In fact, in the *Laws*, written close to the time of the *Timaeus*, Plato proposes a Diopeithes-like statute against impiety—especially of the type produced by men who think that "the whole of the heavens and everything pertaining to it has been generated . . . not by intelligence . . . nor by a God, nor by art, but by what we are talking about; by nature and by chance."[16] Unlike such men, Plato explains that the cosmos as *we* experience it is the artistic creation of a God. Plato's God is of the new breed produced by men like Anaximander, however, and this is extremely important. There is nothing whimsical or arbitrary about this God's actions in creating the universe, nor in his (indirect) governance of it. Every divine act of creation can be directly inferred from God's goodness and rationality: thus the created universe is at once moral and intelligible. Furthermore, there is in it no room for arbitrary and willful divine interventions. Its fundamental rules are fixed once and for all at the beginning, and though God's governance continues through the agency of secondary deities (demons or angels, depending on one's biases), even that governance is by general rule rather than by particular intervention.

The basic problem faced by Plato is to explain the obvious coherence and perceived pattern of the universe of sensible experience, given his presumption that only the suprasensible world of intelligible forms is truly regular and completely knowable. "Whatever is conceived by opinion and with the help of sensation," Plato tells us, "is always in a process of becoming and perishing[17] and everything that becomes must be created by some cause. This cause or creator of the universe, the Demiurge," Plato says, "is hard to find out,"[18] but we do learn something of him through his creation.

First, one asks why God or the Demiurge created the universe at all. God is good, and what is good cannot be jealous; therefore, "he desired that all things be as like himself as they could be. This is in the truest sense the origin of creation and of the world. . . . God desired that all things should be good, and nothing bad, so far as this was attainable."[19]

Existence is better than nonexistence, so God created the universe to be as much like the realm of ideas (i.e., of pure existence) as anything sensible could be. Thus the world is an image, or copy, of the eternal ideas; but because the sensible can not be eternal, God created time as a moving image of eternity.[20] Order is better than disorder, so God created an orderly cosmos. Intelligence, which presupposes soul, is better than its absence, so God endowed the creation with intelligence and soul. Life is better than nonlife, so God created the world as a living being. To contain everything is better than to lack something, so God created the universe to contain all things, and therefore there is but one universe. Finally, to be self-sufficient is better than to be dependent, so God created the world without needs; and because it lacked nothing and had no need for such things as arms or eyes or legs, it was created in spherical form and given the rotational motion proper to that form.

Plato asserts that in order to be visible the universe must contain fire, and in order to be tangible it must contain earth; that is, it must involve something that produces light and something that has solidity. But, he argues in a Pythagorean vein, there must be something to mediate between these two, and in a three-dimensional universe there must be two mean proportionals between the extremes. Thus, in addition to fire and earth, Plato adds air and water as the primary elements of the body of the world. Plato gives the above introductory overview of the creation and constituents of the world in the first few pages of Timaeus' first major speeches, then he returns to fill in the details, only a few of which we can consider here.

The first major stage of the creation is production of the world soul to rule and govern the universe.[21] Mixing together portions that are indivisible, unchangeable, and purely rational with portions that are material, God creates a third entity to bind rationality and materiality together. This stuff he slices in a complex way and disposes into two bands—the celestial equator and the ecliptic (the "same" and the "different")—encircling the world. To the band of the same is given authority over the daily rotations of the heavens and events that occur with uniformity. The band of the different is tilted with respect to that of the same and split into seven bands, each of which governs the motions of the sun, the moon, or one of the known planets. Next

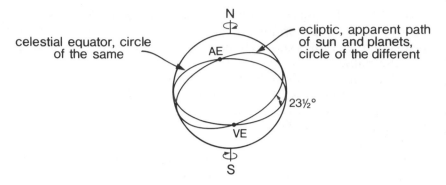

FIG. 7.   Circles of the same and the different
(the celestial Equator and the ecliptic).

he creates the bodies of the stars, sun, moon, and planets, and lights
a fire in the sun to illuminate the whole, creating night and day. Thus
are explained all heavenly phenomena, including planetary retrograde motions, eclipse phenomena, the different periods of revolution of the planets, and so on. Finally, God promises the heavenly
bodies that they will be indissoluble, for "only an evil being would
wish to undo that which is harmonious and happy."[22]

Now that God has created the immortal bodies of the universe,
he assigns to them his authority to create the bodies of three classes
of mortal beings—plants, animals, and men—for if God had created
them himself they would have been immortal. But God reserved to
himself the creation of the souls of animals; and he insisted that men
be created with sensations, feelings, and reason so that he, God,
would be "guiltless of future evil in any of them."[23] As an example of
one sensation, Plato deals at length with sight, for sight is to him the
most imporant sensation, leading through observation of the heavenly bodies to truly philosophical considerations. In the process of
considering sight Plato digresses to discuss not only the visual
organs but also problems associated with the propagation and reflection of light.[24]

At this point the direction of the *Timaeus* takes a very self-
conscious twist. Up to now, Plato tells us, he has been discussing
almost exclusively the rational side of the creation, but he *has*
warned us that the universe was only going to be as good as it *could*
be, not perfect. Now is the time to discuss the limits imposed upon
the creation caused by the materials God had to work with in the first

place. At first the material is "formless and free from the impress of any of those shapes which it is hereafter to receive from without."[25] By this statement Plato does not seem to mean that earth, air, fire, and water did not exist, only that they were so chaotically mixed as to be unrecognizable.[26] Creation, then, involves sorting out the pre-existing elements into coherent patterns. But since those patterns must depend upon the initial elements Plato embarks on a lengthy discussion of earth, air, fire, and water, and their combinations and motions.[27]

Basically, fire, air, and water are different spatial arrangements of 30°−60° right triangles. Fire is composed of regular tetrahedra whose faces each contain 2, 6, or 8 triangles. Air is composed of regular octahedra whose faces contain 2, 6, or 8 triangles, and water corpuscles are regular icosahedra whose faces contain 2, 6, or 12 triangles. Since all three elements are composed of the same basic triangular elements—fire of 8, 24, or 48 triangles; air of 16, 48, or 96 triangles; and water of 40, 120, or 240 triangles—each of these elements can be converted into either of the others. One first-level water corpuscle (40 triangles) could, for example, produce a second-level fire corpuscle (24 triangles) and a first-level air corpuscle. Earth corpuscles are cubes, formed of faces of 2, 4, or 8 45° isosceles right triangles and are therefore not transformable into the other elements. But the sharp corners of the other elements may break up a large earth corpuscle to produce 2 second-level, or 4 first-level, corpuscles. In this way, such phenomena as phase changes or solubility of earthy materials can be accounted for. The sterochemistry of Plato's elements is fascinating, but it is of marginal intellectual importance outside of the scientific tradition, so we will not follow it in detail.

After Plato's discussion of the elements in the *Timaeus* he goes on to use the final third of the work to discuss physiological and what we would call psychological phenomena. This final third was un-translated in antiquity, and was not widely known until the twelfth century, so it was of little or no early cultural importance.

Gregory Vlastos has very beautifully summarized the main features of the *Timaeus* and the reason for its long-term cultural significance, which we look at in detail later. He writes,

> The clientele for Democritian materialism could not be large, even among intellectuals. The vast majority found faith in the supernatural a spiritual necessity—many of them still do. For these the *Timaeus* offered a brilliant alternative. If you cannot expunge the supernatural,

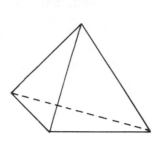

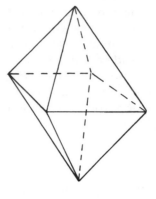

Fire   Air

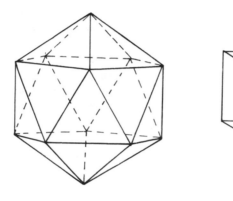

Water   Earth

FIG. 8.   Geometry of Plato's four elements.

you can rationalize it, twining it paradoxically into the very source of the rational order, restricting its operation to a single primordial creative act which ensures that the physical world would not be chaos but cosmos [ordered] forever after. This Plato accomplished by vesting all supernatural power in a Creator who was informed by intelligence and was moved to create our world by his love of beauty and by his pure, unenvying, goodness. Reducing all lesser divinities to creatures, dependencies, and servants of this supreme deity; making the courses of the stars movements of rational agents performing faultlessly their chronometric tasks; grounding all physical causes in the architectonic structure impressed on matter by its divine artificer, Plato, in his own perversely original way, sustained the faith that our world is a cosmos. He gave rational men a pious faith to live by.[28]

IS THE TIMAEUS SCIENTIFIC?

Once again we must pause to reflect. Is it reasonable to say that the *Timaeus* is science—a product of scientific attitudes and activities? Unless the answer to this question is yes, later claims regarding the impact of the *Timaeus* on Christian theology cannot be seen as claims regarding the impact of science upon culture at all.

Surely the *Timaeus* undeniably aims at recognizing the underlying causes of phenomena. In many cases these causes are formulated mathematically and in terms of spatial relations—common scientific explanatory causes even today. But in some cases they are expressed in terms that are moral and aesthetic—as when the universe is assigned a spherical shape because it has no need of appendages, or when the universe is assigned a soul because intelligence is deemed better than its absence. Can such nonempirical judgments be allowed in something for which the name of science is to be claimed? For better or worse, virtually all scientific theories utilize some such untestable presuppositions. Modern demands that the laws of physics be invariant under certain mathematical transformations are such assumptions; and the founding assumptions of modern geology—that only processes that are currently observable may be used to account for historical events—is such an assumption. The assumptions of modern physics and uniformitarian geology have far greater plausibility to us, in part because the explanatory schemes that use them have been extremely successful, but they serve precisely the same theoretical functions as Plato's demands for economy of form and for inclusiveness of desirable attributes.

What of Plato's appeal to the supernatural agency of the Demiurge? On the face of it this appeal seems to disqualify the *Timaeus* from the category of science. Yet many nineteenth- and twentieth-century scientists would willingly admit the original creative role of a deity, and any number of seventeenth-century scientists, including Newton, spoke of God's activity in their scientific works. The crucial issue here is that emphasized by Vlastos—Plato's God is simply the name for that which imposes order and structure upon the universe, and even those acts done by the Demiurge are given *reasons*.

But does Plato seek universally valid knowledge? Surely some thing that is admittedly nothing but a "likely story" doesn't meet the criteria established for science. Today almost all scientists agree that scientific theories are all provisional—to be accepted only until new information demands revision. It is generally assumed that scientists use models that are in some sense mere "likely stories" whose special virtue is that they have internal coherence, and that they are in conformity with the set of observations deemed relevant at the time. Thus, the *Timaeus* clearly seems to meet any reasonable requirements of a scientific document. Furthermore, it certainly involved an intelligent and self-conscious use of prior astronomical, mathematical, and physical thinking, and was thus a distinct part of a scientific tradition.

PLATONIC ORIGINS OF THE CONFLICT BETWEEN SCIENCE AND POETRY

Today the most vitriolic and consistent opposition to scientistic and scientific attitudes is voiced by professional poets. Though slightly more bitter than most, e. e. cummings expressed a relatively common twentieth-century poetic response to science in the following letter to his friend Eva Hesse:

> It might be said that like Death, S is fundamentally a depersonalizing leveler (47) whereas I stand for individuality and personal uniqueness as against sameness or standardization (31–32)—that, so far as I am concerned, mystery is the root and blossom of eternal verities (11, 43, 82, 110) while, from a scientific standpoint—mystery is something to be abolished at any cost—that for me nothing impersonal or measurable matters (68, 110) but for science measurability and impersonality are everything—and finally that to know anything equals (by my values) to be merely undead; whereas "to feel something is to be alive."[29]

In today's culture the scientist is largely indifferent rather than antagonistic to the poet because his position of centrality in the culture is so solid that the poet's quaint attitudes offer no significant threat or challenge. In Hellenic antiquity, however, the social roles of science and poetry were nearly reversed. The poet's specialty was perceived as absolutely central to civic life, for through poetry—especially through Homeric and Hesiodic epic and through tragic drama—Greek society educated its young and held forth exemplars to illustrate and inculcate societal values. With rare exceptions, like those expressed in Aristophanic comedy, the poets of Hellenic Greece ignored the rise of scientific throught. But Plato, writing in support of the new scientific attitudes, realized that there was a fundamental conflict between the poets' claims to be the educators of Greek society and his own claims for a very different kind of education for virtue. Thus Plato launched a bitter attack on poetry in the name of science (or philosophy), initiating discussion of many of the issues that still inform arguments between spokesmen for scientific and humanistic ways of responding to the world.

It is very hard to see Plato as an antipoetic scientist, for Plato was himself a poet, communicating his ideas in allegorical and mythic form. But in Book Ten of the *Republic* Plato attacked poetry and banished it from his ideal state, and as he continued to write, his works became less poetic until in the *Parmenides* and *Laws* there is nothing left but expository prose.

Plato's fundamental reasons for opposing poetry have been extensively discussed by Eric Havelock who summarizes the issues as follows:

> There was a state of mind which we shall conveniently label the "poetic" or "Homeric" or "oral" state of mind, which constituted the chief obstacle to scientific rationalism, to the use of analysis, to the classification of experience, to its rearrangement in sequence of cause and effect. That is why the poetic state of mind is for Plato the arch-enemy, and it is easy to see why he considered the enemy so formidable. He is entering the lists against centuries of habituation in rhythmic memorized experience. He asks of men that they should *think* about what they say, instead of just saying it. And they should separate themselves from it instead of identifying with it; they themselves should become the "subject" who stands apart from the "object" and reconsiders it and evaluates it, instead of just "imitating" it.[30]

Greek poetry had provided the method of teaching successive generations of Greeks customs and values that constituted public and private law and morality, and engendered piety and patriotism. The Hellenic Greek "learned" by becoming emotionally involved with (i.e., by identifying with so that he would relive or imitate) the heroes of epic poetry or tragedy. He did not learn ethics, politics, and other skills through rational analysis. In fact, the less poetry encouraged men to *think* about their actions the more effective it was, and the more it simply conditioned them to respond unhesitatingly to circumstances they faced. It was just this anti-self-reflective charac-

teristic of the teaching function of traditional poetry that Aristophanes lauded in the Frogs. And it was just this characteristic that Plato
had to oppose, given his belief that knowledge could be had only of
the world of ideas and not of the fluctuating world of sensory
experience. How could one possibly learn by "imitating" something
or somebody that was nothing but part of the imperfect sensory
world? Even if they chose the best possible examples, the poets could
only sing of flesh and blood heroes or of the acts of the gods in the
world of human drama. They could never convey true knowledge of
stable ideas; they could but offer opinion. Thus those who learned
from the poets could never attain real knowledge:

> Those who always want to hear something new—[who] run about to all
> the Dionysiac festivals, never missing one, either in the towns or in the
> country villages. . . . the lovers of sounds and sights [who] delight in
> beautiful colors and shapes and in everything that art fashions out of
> these—[are, nonetheless] incapable of apprehending the nature of the
> beautiful in itself.[31]

But poetry was not harmful just because it distracted humans
from Plato's true method of learning. It was positively evil in many
ways. First, as Plato points out in Book Three of the Republic and as
Xenophanes had long held, the models held out for imitation by the
poets were often far from the best possible ones, and the habit of
imitating inferior models could have a terrible impact upon the
character formation of impressionable youth.[32] The most important
criticism of poetry, however, depends upon Plato's tripartite division
of the soul into rational-calculative, irrational-appetitive, and high-
spirited parts—that is, one part that "reckons and reasons," one part
that "loves, hungers, thirsts, and feels the flutter and titillation of
other desires," and one part that provides courage and bravery.[33]

According to Plato there is no question that the rational-
calculative part of the soul should be the master and ruler over the
other two parts.[34] But poetry appeals almost exclusively to the
irrational-appetitive part of the soul, making it the master over that to
which it is properly subordinate. Since there is no theme in the
Republic more central and repetitive than that good consists in
having the various parts of any entity fulfill their proper functions in
harmony, the poetic tendency to make emotion dominate reason is
totally unacceptable. Thus the chief argument against poetry is its
power to corrupt even the better sort of men, making them slaves to
their desires and loosing the constraints of reason.[35]

Plato concludes,

> We can admit no poetry into our city save only hymns to the Gods. . . .
> For if you grant admission to the honeyed Muse in lyric or epic, plea-
> sure and pain will be lords of your city instead of law and that which
> unfolds itself to the reason as best.[36]

So does Plato provide the scientific counterpoint to Aristophanes' praise of unreflective poetry, and sound the basic themes
which e. e. cummings finds anathema.

For Plato, only ideas that are divorced from the sometimes
good, sometimes bad characteristics of living humans are of real
interest. For cummings and for most of the poetic tradition, the
personal, individual, and unique holds the greatest value. For Plato,
mystery is simply an indication of failure to understand—something
that calls out for transformation into knowledge. Whereas for cummings and the poets, mystery is the "root and blossom of eternal
verities," something to be retained, to save a sense of awe and magic.
For Plato the paradigm of knowledge is quantification—what is
measurable is real; but to cummings, "nothing measurable matters."
Finally, for cummings and the poets, "to know anything is to be
merely undead, whereas to feel something is to be alive"; but for
Plato and for the scientific tradition, to merely feel is to be a slave to
desires, while to know is to become their master.

THE TRANSITION FROM HELLENIC
TO HELLENISTIC CULTURE: THE
ARISTOTELIAN INTELLECTUAL EDIFICE

Plato was undoubtedly the most important Greek intellectual for the
development of Western culture between about 300 B.C. and A.D.
1200. For example, his philosophical perspective most directly
informed early Christian thought and doctrine—to the extent that
Christian doctrine was philosophical at all.

But Plato's ideas were not unchallenged. Within less than
thirty-five years of his death there were at least four major intellectual
schools in Athens competing for students, attention, and influence.
All of the important Greek Hellenistic intellectual traditions derived
from these schools. Plato's Academy continued its existence—with
an initially increased emphasis on mathematical studies—for nearly

eight hundred years. Under the leadership of Epicurus (d. ca. 270 B.C.), a school identified as the Garden developed the social and ethical implications of the Democritian materialist tradition. Under Zeno of Cition (d. ca. 264 B.C.), a Stoic school, usually called the Porch, was founded. And under Plato's most famous student, Aristotle (d. ca. 323 B.C.), the Lyceum came into existence.

Of these schools, the Porch encouraged the most immediate and direct impact of scientific attitudes and activities on the broader culture through its incorporation of astral determinism into Stoic doctrines (see chapter 2). And Epicureanism (the Garden) produced a fascinating, short-lived atheistic movement. But the greatest long-term impact was from Aristotle's Lyceum. Aristotle and his students produced not only the most extensive and self-conscious ancient discussions of logic and scientific method, but also the most inclusive ancient system of scientific analyses, extending from metaphysical through physical, astronomical, meteorological, biological, psychological, ethical, political, rhetorical, and even poetic topics. Though temporarily eclipsed by Stoic and Platonic doctrines in antiquity, Aristotelianism survived—even thrived—at the Alexandrian Museum from about 300 B.C. to about A.D. 300. It was carried into Syria by the heretical Nestorian Christians, became a central feature of Islamic intellectual concerns during the period of Stoic and Platonic supremacy in Europe, and reemerged as the dominant framework for European intellectual life in the thirteenth century, when it was reintroduced through translations from the Arabic.

A full treatment of Aristotelian thought is far beyond the scope of this book. But some central ideas must be presented to illustrate the most elaborate Greek extension of earlier scientific ideas, and to provide a grounding for discussions of medieval theology and of Renaissance attitudes toward the relationships between theoretical science and practical activities.

Unlike Plato, who seldom acknowledged his intellectual debts in order to claim uniqueness for his own ideas and those of his teacher Socrates, Aristotle not only admitted his debts to earlier thinkers, he self-consciously explored them. In fact, Aristotle insisted upon the fact that his own attempts to understand such disparate subjects as metaphysics and politics were the direct outgrowth of an intellectual tradition that began when the Milesian naturalists (in Aristotle's terms, the *physicoi*) separated themselves from the traditional poetic mythographers (the *theologoi*) by seeking new naturalistic and rational explanations of astronomical, meteorological, and geological phenomena.[37] For him, as for Adam Smith (see chapter 1), speculation about the physical world led progressively and directly to questions about how knowledge in general is possible, and then to questions about how and if the proper means of knowing suggested by physical speculations could be applied to questions of practical sociopolitical concerns (i.e., to ethics, economics, and politics), and to problems associated with human creations (i.e., to technology, medicine, and even the writing of drama).

According to those who became admirers of Aristotle, his greatest accomplishments were (1) the development and codification of a universally applicable scientific method, (2) the classification of sciences in such a way as to delineate the extent and limits of each, and (3) the initiation of a systematic process of information collection and analysis, which provided the basic foundations of almost all major sciences. Aristotle's descriptive work in animal biology was most thorough and successful by our twentieth-century standards; but his attitudes toward the relations between theory and practice, his general categories of analysis and classification, his discussions of scientific method, and his cosmology were to have the greatest impact upon Western culture.

ARISTOTLE ON THE RELATIONSHIP BETWEEN THEORETICAL AND PRACTICAL SCIENCES

In order to understand a major difference Aristotle saw between kinds of knowledge—that is, the distinction between productive, practical, and theoretical sciences—we must talk about the ends that different kinds of knowledge might serve. We begin by briefly inspecting one of Plato's attitudes. So far as we can tell, Plato at least *began* seeking knowledge principally for practical purposes—above all for the improvement of life in the polis. In the process of philosophizing, however, Plato discovered such intense satisfaction that he came to see the search for knowledge as perhaps the highest source of individual happiness—a source so seductive that it threatened to draw the best men away from their obligation to serve the state and to improve the quality of life of the polis as a whole. Thus, in the *Republic*, Plato frequently comments on the need to force those who have come to know The Good to return to help their fellow citizens. And he acknowledges that in some sense, to "compel them to add to their duties the care and guardianship of the other people" really is to "make them live badly, when they might live better."[38]

In Aristotle the Platonic tendency to elevate the life of intellect, study, and contemplation become much more extreme and overt. The first book of Aristotle's *Metaphysics*, or first philosophy, begins with the assertion that "all mankind have an instinctive desire for knowledge."[39] He goes on to claim that men first came to study philosophy out of a sense of wonder and for the sake of knowledge itself rather than for any utilitarian purpose.[40] For this reason, "among the various sciences, that which is pursued for its own sake and with a view to knowledge has a better claim to be considered wisdom than that which is pursued for its applications."[41] In the *Metaphysics* Aristotle briefly struggles, not simply to *assert* the dignity of pure knowledge but also to explain *why* it should be desired beyond knowledge that is directed to producing objects, or to practical social activities. Pure knowledge is a knowledge of the ends toward which things are done and not merely of the means. Thus, it is a knowledge that commands the subordinate knowledge of mere techniques; and surely it is better to be one who commands rather than one who must accept the commands of another.[42]

Aristotle produces a much more extended and compelling justification of the disinterested search for pure knowledge in the *Nicomachean Ethics*. Here he builds upon a theory of the hierarchy of souls in living beings, just as Plato had justified his antagonism to poetry with a theory grounded upon a similar hierarchy. For Aristotle as for Plato the greatest goods in life arise out of the fullest exercise of the highest capacities that an entity or organism possesses. In a trivial way this is why we call the racehorse that runs faster than any other the best horse, for it most fully manifests its capacity for speed. The capabilities of living things are defined for Aristotle by what we translate as "soul." All organisms have a vegetative soul, which allows them to grow by somehow converting nourishment into tissues, and which provides for the possibility of procreation. All *animals* have a sensitive or appetitive soul, which allows them to identify and to seek out things that will fulfill their bodily needs. But humans alone possess a rational, or intellectual, soul.[43] And because man alone possesses intellect (i.e., the ability to recognize the causes and reasons for things) man's greatest good must be related to the manifestation of that ability.

So far, Aristotle's doctrine of soul differs little from that of Plato. But Aristotle further distinguishes between two aspects of our rational capacities. "With one," he says, "we apprehend the realities whose fundamental principles do not admit of being other than they are [such realities will be defined as the subjects of the theoretical sciences, since they are beyond our capabilities to alter them in any way], and with the other we apprehend things that do admit of being other than they are [such realities will be the subject of practical sciences such as ethics and politics, because they lie within our capacity to influence them]."[44] Since both the capacity to apprehend theoretical sciences and the capacity to understand practical sciences are uniquely human, both are associated with human goods. Moreover, although the end of theoretical knowledge is contemplation rather than action and the end of practical knowledge is good or just action, both contemplation and action in this sense are ends in themselves; so neither theoretical nor practical science is subordinate to other ends (at least in the first instance), as merely productive knowledge, such as carpentry, is.

Yet, Aristotle claims that contemplation is a higher end than action for three critical reasons: (1) it is more self-sufficient, (2) it is more leisurely, and (3) it is more nearly divine. First, Aristotle asserts that self-sufficience is a value in its own right, and that contemplation is self-sufficient in a way that practical action is not because action depends on the behavior of others.

> The just man still needs people toward whom and in company with which to act justly, and the same is true of the self-controlled man, and a courageous man, and all the rest. But a wise man is able to study even by himself, and the wiser he is, the more he is able to do it.[45]

Second, like a good Greek aristocrat, Aristotle holds that leisure is good in itself, and that whereas the actions associated with practical knowledge—political and military pursuits—detract from leisure, the study of theoretical sciences is a leisure activity.

Finally, Aristotle makes a claim that was to have tremendous importance for medieval Christian attitudes toward the value of contemplation as opposed to that of action. Man has a spark of the divine in him according to Aristotle, just as he was created in the image of God according to the Judeo-Christian tradition. This divine element, "though it is a small portion [of our nature], far surpasses everything else in power and value—it is the controlling and better part."[46]

For reasons associated with the scientific impact on Greek religious ideas (see chapter 3) Aristotle was forced to claim that "the activity of the divinity which surpasses all others in bliss must be the contemplative activity."[47] The human activity that is most divine, then, must be contemplation, and man's greatest happiness must be in contemplation.

> The Gods enjoy a life blessed in its entirety; men enjoy it to the extent that they attain something resembling the divine activity; but none of the other beings can be happy, because they have no share at all in contemplation or study. So happiness is coextensive with study, and the greater the opportunity for studying, the greater the happiness, not as an incidental effect, but as inherent in study.[48]

Because contemplation is the highest function of man, practical philosophy is ultimately justified because it provides the conditions under which the theoretical sciences—metaphysics, physics, astronomy—can be pursued by a leisured intellectual elite.

> Practical wisdom has no authority over theoretical wisdom or the better part of our soul, any more than the art of medicine has authority over health. [Just as medicine does not use health, but makes the provisions to secure it, so] practical wisdom does not use theoretical wisdom but makes the provisions to secure it.[49]

Scientific activities and attitudes have thus been transformed by Aristotle from the best *means* for achieving the Good life, which they had been for Plato, into the supreme end—the very pinnacle of a hierarchy of human goods.

THE CHARACTER AND CATEGORIES OF ARISTOTELIAN SCIENCE

Like Plato, Aristotle insisted that scientific knowledge of any subject involves a combination of "true opinion" and "logos." We might say it involves both recognition of the *facts* relevant to the subject and an understanding of why those facts are related to one another as they are and in no other way.[50] Furthermore, Aristotle, like Plato, explicitly insisted that scientific knowledge must be universal knowledge in all of the senses discussed in chapter 1. That is, it is the same for all men, for all time, and in all places.

Plato had despaired of discovering such knowledge of anything other than a timeless perfect world of Ideas or Forms, which stands outside of and above the imperfect universe of sensory experience.

But Aristotle took a fundamentally different approach to the character of what can be known scientifically. His attitude was derived from the medical tradition into which he was born and the Democratian empiricist tradition, both of which claimed the centrality of direct sensory experience in determining the character of reality. For Aristotle, the facts that provide the subject matter of scientific knowledge are immediately experienced entities in the sensory world, or characteristics of those entities directly abstracted from experience. A mathematical figure, for example, is a proper subject of mathematical science. This figure is that portion of the experience of some physical entity remaining when complicating features such as color, texture, heaviness, etc. are ignored. When this has been done, we say that the mathematical figure has been abstracted from our experience.[51]

Now, one of the primary characteristics of our immediate experiences is their particularity. We experience a specific dog or a specific tree rather than "dog" in general, or "tree" in general. Each such specific object of experience is designated as a "primary substance," and Aristotle argues that the most fundamental reality is the existence of primary substances, which can be experienced.[52] Two things about these primary substances and the way in which men interact with them are of great importance to the generation of scientific knowledge.

First, if it were not the case that in nature different primary substances were related to one another in certain special regular ways, there could be no science. Science must be about universals—that is, scientific statements must apply to more than a single primary substance. Science is possible, then, only because there are classes of primary substances that share salient features, as the class of man is composed of primary substances that share the abilities to walk on two legs, to speak, and to reason. Such classes, or "secondary substances," are the real subjects of scientific discourse. But sciences could not exist even in the presence of natural classes of substances except for the fact that humans are so constituted that they can recognize the relevant similarities among primary substances. Features of a particular experience are retained in memory. When we experience another substance of the same class we recognize the similarities, and repeated experiences fix in our minds the existence of the class of substances identifiable with one another because of their shared characteristics.[53] Aristotle uses the term "form" to denote the necessary attributes of members of a given class or secondary substance, but he insists that forms do not exist apart from the primary substances of our experience, as Plato had presumed.

For Aristotle, the problem of deciding what form is leads to a discussion of three fundamental dichotomies that underlie all of his scientific discussions. First is the matter-form dichotomy; second, the substance-accident dichotomy; and third, the potentiality-actualization dichotomy.

One way of analyzing or thinking about a substance is to distinguish in our mind between the properties or charcteristics that determine the kind of thing it is (the class to which it belongs) and what "stuff" it is that those properties make into the thing. The first is the form of the substance, and the second is the matter.[54]

There is never, properly speaking, form without matter or matter without form. But the same matter is, in principle, capable of taking on indefinitely many different forms, and this fact provides Aristotle with a means to understand many natural phenomena by appealing to a second distinction—between potentiality and actualization. Consider the following simple example: We wish to explain or understand the change of an acorn into an oak tree. Aristotle tells us that the acorn can be thought of as a piece of informed matter. And part of the form of an acorn is that it presents the matter that it informs with the potential to take one and only one subsequent form, that of an oak tree. Because of its form, the acorn cannot turn into an ash tree or a snake, or a man, but it can change into an oak; so we can say that the growth of an oak tree is the process of actualizing an acorn's potential to be an oak.

One kind of change thus occurs when the same matter changes form, guided by the potential inherent in its initial state. Not all changes, of course, are changes of form, nor does the realization that some particular change is a *transformation* provide an adequate or complete explanation of that change. But the actualization of a potential new form constitutes an interesting and important kind of change; and recognizing that something is a transformation is a necessary part of the complete explanation of such changes.

It is important to note that although some transformations seem to be completely spontaneous, it is often the case that they do not take place except through the agency of some obvious external influence. The potential that a woman has to be pregnant, for example, can only be actualized through the male's act of impregnation. Even when the agent, or efficient cause, of a change is not obvious Aristotle insists that there must be some agency involved, so he suggests that sometimes that agent is provided by the initial form itself.

By way of example and for future reference, we briefly consider a particular set of transformations that terrestrial substances (substances that exist on, in, or near the Earth) may undergo, and then compare them with changes experienced by substances that lie beyond the sphere of the moon and provide the subject matter of astronomy. The most elementary kinds of informed matter that exist on or near the earth we designate as the Aristotelian elements, earth, water, air, and fire.[55] For reasons too complicated to analyze here, Aristotle concludes that the form of earth is dry and cold, the form of water is moist and cold, the form of air is moist and hot, and the form of fire is hot and dry. Furthermore, it is the nature of what is dry that it is potentially wet and vice versa, and it is the nature of what is cold that it is potentially hot and vice versa. Thus, subject to the proper conditions (not yet specified) it is possible for any given terrestrial element to be transformed into any other. This character of the terrestrial elements helps to explain such phenomena as combustion (the transformation of earthy materials into fire) and the formation of rain in the atmosphere (the transformation of air into water).

In contrast to the forms of terrestrial elements, the form of

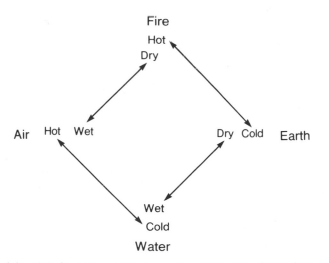

FIG. 9.   Aristotle's four terrestrial elements and modes of transformation.

heavenly substances does not involve opposing qualities that are transformable into one another; so heavenly bodies are eternal.[56]

But not all changes are transformations; to understand other kinds of changes we must introduce yet another way of thinking about substances and their characteristics. While substance is the most basic mode of being, Aristotle says that there are other modes of being, or *accidents* that inhere in substances. When we consider a piece of grass, for example, we may think of it as green (an accident of quality), as being produced from a certain seed (an accident of relation), as being eight inches long (an accident of quantity), and as being in a certain field (an accident of place).[57] No set of accidents can define the form of a substance, for many different accidents may belong to one kind of thing. Rather, it is the form of a substance that accounts for the range of accidents that the substance may have.

Once again this fact helps to explain certain kinds of phenomena. For example, one of the characteristics of matter that has the form of any physical object is that it may be located in one place and have the potential to be in some other place. What we call motion, or what Aristotle called local motion, can thus be understood as a process of actualizing a body's potential to be in some place where it was not initially. In such a motion the form stays constant but there is a change in the accident of place. The form acts only to specify the range of possible places to which the body might go.

To understand why this notion of formal restrictions on places is important we can consider the distinctions between terrestrial and heavenly bodies and their local motions. Just as it is part of the form of earth to be dry and cold it is part of the form of earth to naturally fall toward the center of the universe (which thus incidentally becomes the center of our Earth). It is part of the form of fire to rise from the center toward the sphere of the moon, and it is part of the forms of water and air to seek their natural places in concentric rings at successively greater distances from the center of the universe and above the earthy element.[58] These characteristics of the terrestrial elements help to explain, for example, why, when wood burns, the fire and smoke rise; and why, when raindrops come to exist within the airy atmosphere, they drop toward the ground. No change of form occurs in these local motions, but the local motions involve changes in the accident of place, which are directed by the forms of different kinds of matter.

Unlike the forms of terrestrial elements, which naturally produce motions toward and away from the center of the universe, the

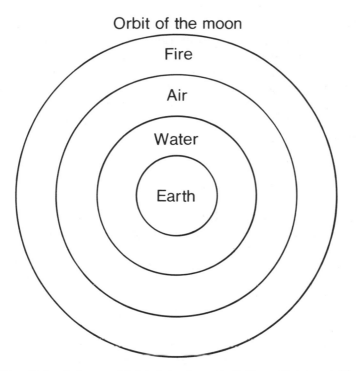

FIG. 10.   Natural places of Aristotle's elements in the sublunary world.

form of heavenly matter allows no potential for increasing or de-
creasing the distance from the center. Instead, the form of heavenly
matter—the aether—allows only for motions that remain equidistant
from the center; thus the heavenly bodies move only in circular paths
above the sphere of the moon.[59]

There is one final characteristic of the world as Aristotle sought
to understand it that we must consider before we explicitly consider
his pronouncements on scientific method. Both his training as a
student of Plato and his interest in living organisms led Aristotle to
insist that no adequate knowledge of the world and its contents could
exist unless it involved a notion of purpose, or the ends toward which
any given process or structure was aimed. This concern with pur-
posiveness is what most clearly distinguishes Aristotle's scientific
attitude and activities from those of most twentieth-century scientists. Strange as they may initially seem, Aristotle's categories of
matter and form are easily developed into later concepts of matter

and "laws of nature" (which Francis Bacon continued to call forms). His notion of accidents is easily transformed into our notion of the observable properties of objects. And his notions of potentiality, actuality and transformation are not radically different from those used by modern geneticists when they talk about the way in which the potential behaviors coded in genetic material are only partially actualized by the limited environmental stimuli which any given organism encounters, or by modern physicists when they speak of the transformation of mass into energy or of one form of energy (i.e., heat) into another (i.e., electricity). We may think that Aristotle made terrible mistakes in his attempts to employ his categories. And we may not often use his ways of thinking, but we must acknowledge that Aristotle's methods of approaching the world in these terms were of central importance in giving shape to modern scientific attitudes and activities.

Except for certain limited technical applications in biology, Aristotle's notions of purpose, or teleology, have been totally renounced among modern scientists; yet they were of prime importance to Aristotle, and to his reception in the Middle Ages. It seemed patently clear to Aristotle, as it had been to Plato, that no activity, natural or human, was undertaken without some end in view. While this feature of activity might have been most obvious in connection with the human productive arts—no carpenter would undertake to build ships, for example, unless someone desired them to use for certain commercial or military purposes—it seemed almost equally apparent in connection with living beings whose structures were obviously related to the needs of organisms. Eyes, for example, surely seemed to arise in response to organism's needs to recognize both objects of desire (like food) and of aversion (like enemies). Teeth clearly were provided for the purpose of tearing or grinding food. Even social institutions were clearly developed to serve certain collective needs. Thus it was natural to suppose that all phenomena should be analyzable in terms of the ends they served.

In the realm of inorganic nature the demand for ends in view produced results that may seem bizarre and incorrect from a twentieth-century perspective. Aristotle, for example, entertained the possibility that once a physical body is set in motion it might simply continue to move in a straight line indefinitely until interfered with in some way[60] (we call this the law of inertia) but he was forced to deny that this rule could possibly hold; there could be no determinate end to such a process. Natural motions must not simply be away from some initial place, they must be toward some specific final place and for some specifiable purpose.

Most of the purposes specified by Aristotle are what we might call functional. That is, they have to do with serving ends like procreation, acquiring food, avoiding enemies, and so forth. But in some cases—most notably in connection with his theory of heavenly motion—Aristotle appealed to such purposes as "the love of the divine." Such ends had particular appeal to Christian scholars of the Middle Ages.

Reflecting briefly on the factors that Aristotle used to explain events and objects in the world of substances, we discover that he explained the why of things in four complementary ways—in terms of their matter, their form, the active agents involved, and the purposes or ends in view. Aristotle summarized this feature of his scientific theorizing by stating that we completely understand a phenomenon when we can specify its four *causes*—its material cause, its formal cause, its efficient cause (the active agency), and its *final* cause (the purpose of the thing or event).[61]

ARISTOTLE'S METHOD

So far, we have seen that Aristotle's scientific theories emphasize *phenomena*; they explicitly emphasize the fact that scientific knowledge is *universal*; and they seek to recognize the *underlying causes* of events and objects. In chapter 1 we also claimed that scientific knowledge was *systematized* knowledge. It is once again to Aristotle that we owe the most extensive and explicit ancient discussion of this feature of science. In fact, of all of Aristotle's writings, his discussions of scientific method and classification alone (carried out in works entitled the *Categories, Topics, Prior Analytics, Posterior Analytics,* and *Interpretations*) were widely studied throughout antiquity and never totally lost. Of these, the first two were translated into Latin in the sixth century by Boethius to become a standard part of medieval education. It was not until the late twelfth and early thirteenth centuries that the remainder of his works were recovered and studied.

We already know that Aristotelian science sought to understand both the facts connected with a variety of topics and the reasons why those facts had to be as they were. Moreover, we know that the reasons have to be related to the four kinds of causes operating in the world. But we do not yet know how scientific knowledge can be built into a *demonstrative* system linking known facts and causes to one another.

Aristotle, unlike Plato, did not believe that the subject matter of mathematics should influence the conceptual content of all other sciences; but like Plato he did believe that the structure of mathematical knowledge should serve as a model of the structure of all other sciences, because mathematics alone provided a model of logical proof of statements about a subject matter. While one might, and usually did, discover true statements about the world through sensory experience, one could not claim *scientific* knowledge until those statements could be demonstrated to be not simply true but also necessary.

The process of constructing a science, then, involved discovering truths and demonstrating those truths from a set of first principles, which was not demonstrable from within the structure of that science itself.[62] In order to carry out this process, two fundamental problems had to be addressed. First, a technique for deriving necessary conclusions from assumed truths had to be developed. Such a logical technique, the syllogism, Aristotle invented in the *Prior Analytics*.[63] Second, some method for establishing the "unproven" first principles from which to argue syllogistically had to be established.

For present purposes we need only recognize that the syllogism is a general technique that produces necessary conclusions when two statements (premises) are initially accepted as true. Only the first figure of the syllogism is of central scientific importance,[64] so we will limit our considerations to it. A general statement of the first figure appears below on the left; a typical example appears on the right.

| All A are B | All stones are earthy bodies |
All B are C	All earthy bodies fall to the ground
All A are C	All stones fall to the ground

Note that in the example the conclusion does not really tell us something that we did not know already. That stones fall is generally discovered experientially. What the syllogism does is to *add* to our awareness of the truth of the conclusion a demonstration of *why* it is true. In that sense the premises of the syllogism are the causes of the conclusion; and as causes, the premises of syllogisms involved in science are always universal, and always tell us something about the matter, form, efficient cause, or final cause of the entity or process treated in the conclusion. The syllogism of the example is, therefore, a properly scientific syllogism, since its premises are universal statements detailing the material and formal causes of the fall of stones.

In the *Posterior Analytics* Aristotle addressed the critical problem of how to establish the first principles of a science—that is, the premises of the most fundamental syllogisms—in the first place. Since they are prior to any syllogisms, the first principles cannot be the product of scientific reasoning. Some first principles are common to all sciences. Such common principles, like "if equals be taken from equals, the results will be equals" or "the whole is greater than any of its parts," are simply and immediately recognized to be true by any person who reflects upon their meaning. They can neither be proved nor doubted.

The remainder of the first principles must contain at least some that are specific to the particular science under consideration; and these again must be of two kinds.[65] First, there must be some definitions of the entities constituting the subject of the science. The statements that physics deals with beings "formed by nature"[66] and that motion is "the actualization of what exists potentially, in so far as it exists potentially,"[67] are both such definitional statements. Much effort may go into making the definitions relate to natural categories, but the truth of such statements is established by convention. The definition simply directs us how to use certain terms.

The second kind of specific first principles of a science must state relations between two or more previously defined terms of the science. For example, the statement that all bodies formed by nature contain within themselves a beginning of movement, or tendency toward change goes beyond the definitions of natural body and of motion to claim that the two are related in such a way that the former spontaneously undergo the latter.[68]

The crucial remaining question is how the relationship between natural bodies and motion—or any other such primary relationship—can initially be discovered, and be known to be an unprovable first principle of a science. Though Aristotle's answers to this double-pronged question are not very satisfying, they nonetheless remain even today the foundations of our notions of scientific investigation. Put most simply, the first principles are discovered by observation and induction in connection with the marvelous ability that man has to recognize constant relationships, or forms, in nature.[69] How they are not merely discovered but discovered to be *true* is a more complicated issue. In one sense their truth is directly intuited or recognized. But this seldom happens immediately and without some effort. In practice, candidates for first principle status, discovered either inductively or by examining "the accepted opinions of the most notable and illustrious men," are subjected to a criticism that involves developing their consequences and bringing to light any contradiction.[70] As a result of this procedure we finally come to "see" or intuit those that are true.

In precisely such a way did men like Galileo and Newton come to formulate the foundations of classical mechanics on the basis of limited, but carefully considered, experience and prior opinion.

ARISTOTLE'S COSMOLOGY

Because the basic features of the Aristotelian universe, as they were established largely in *De Caelo*, were to form the underlying framework for most medieval literature, philosophy, and theological quarrels, some overview of the Aristotelian cosmos is essential for anyone who hopes to understand later Western culture. Many of the key features of the Aristotelian universe have already been developed in the examples used to illustrate Aristotle's conceptual categories and methodological precepts, but it remains to be seen how these notions fit into some more or less integrated scheme.

Basically, the Aristotelian universe is structured as a set of nesting spheres, all centered upon the center of the universe and of the earth. Nearest the center are the spheres of earth, water, air, and fire. It is within these spheres that all fundamental changes take place; that is, in which all growth and decay, or generation and corruption, occurs. Because it is not part of the nature of any of the elements (earth, water, air, and fire) constituting the bodies in this region to move circularly, the material of each of these spheres is basically at rest at its appropriate distance from the center, though portions may naturally move toward and away from the center as substances come into existence not in their natural place.

Beyond these four central spheres are the nesting crystalline (i.e., hard but transparent) spheres that carry and account for the motions of the heavenly bodies—the moon, the sun, the planets, and the "fixed" stars. Technically, each planet needed a set of four or five spheres to account for its apparently irregular motions;[71] but for our purposes, and those of most medieval scholars, we can consider the set as if it were reducible to a single sphere upon which the observed heavenly body was located as a luminous spot. We have already touched on the material and formal causes of heavenly bodies and motions, pointing out that the circularity of heavenly motion and the unchanging nature of heavenly substances derive from these causes. But we must call attention to at least one implication of these causes.

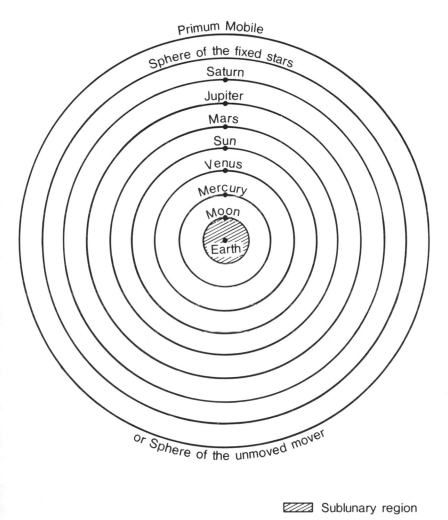

 Sublunary region

FIG. 11.   Aristotelian cosmos.

Heavenly objects, as unchangeable, were of course *eternal*. They could never have been *created* and can never pass away. Nor could the Heavens as a whole be anything but eternal. This implication, of course, provided a substantial problem for Christian scholars, whose canonical book, the *Bible*, explicitly described the *creation* of the heavens as well as the earth.

When Aristotle turned to the question of the efficient causes of heavenly motions, a second serious problem confronted him. At one level he provided a simple and mechanical explanation: the stars are affixed to a hard transparent sphere which rotates once daily. This sphere by its contact with that of the first planet inside carries the latter along, by a kind of friction drive, this sphere in turn drives the next inner sphere, and so on, down to the sphere of the moon. But what is the efficient cause of the motion of the sphere of the fixed stars, and what are the efficient causes of the deviations of the planets from simple diurnal motion?

To answer the first of these questions Aristotle considers the possibility that there is yet another rotating sphere outside that of the fixed stars, which causes the rotation of the sphere of the fixed stars. But he concludes that this cannot be the case; for if we assume an insensible moving sphere beyond the fixed stars, there is no reason to stop asking the question of what moves *it* and answering in the same way. In this way we would be led to an infinite regress. Therefore, says Aristotle, there must be a first and unmoved mover that causes the motion of the fixed stars.[72]

The characteristics that this unmoved mover must have are established in *Metaphysics*, Book 12, chapter 7. It must act solely as the object of desire and of thought, and must therefore be the final as well as the efficient cause of heavenly motion. Furthermore, it must exist eternally and by necessity—without movement, without parts, even without magnitude and place. It is very much like Anaximander's divine aperon. And it is even more like (because it is a prime source of) the later Christian notion of an omnipotent, omniscient, omnipresent, eternal God.

If the efficient and final cause of the motion of the sphere of the fixed stars, or the first heaven, suggests that Aristotle's cosmology is at the same time a theology, his analysis of the efficient and final causes of planetary motions leaves us in no doubt. The spheres that carry the planets are essentially alive and are driven by intelligences whose purpose is to achieve, as far as it is possible, perfection, or identification with the unmoved mover and final cause. This is much easier for the bodies spatially closest to the unmoved mover. Thus, the fixed stars achieve the most nearly perfect possible state—that of a single rapid and uniform circular motion. The planets, being more distant from the divine unmoved mover, strive for perfection by means of the several slower uniform motions that manifest themselves in their apparent wanderings. And the earth, being at the greatest distance of all, is effectively precluded from this kind of heavenly perfection entirely. The whole scheme and its rationale is briefly summarized as follows:

> "To attain the end [of complete actualization] would be best for all; but if that is impossible, a thing gets better and better, the nearer it is to the best. This then is the reason why the earth does not move at all and the bodies near it [the moon and sun] have only a few motions. They do not arrive at the highest, but reach only as far as it is in their power to obtain a share of the divine principle. But the first heaven reaches it immediately by one movement, and the stars [planets] that are between the first heaven and the bodies farthest from it, reach it indeed, but reach it through a number of movements."[73]

Given the radical distinction between terrestrial and heavenly forms and motions, one might reasonably assume that there could be no interaction between the two; but that is not quite correct. Aristotle had to admit what no intelligent naturalist of the fourth century B.C. could deny; that is, that seasonal terrestrial events are in some way dependent on or caused by the sun.[74] In fact, he asserted that in some unspecified way, all coming to be and passing away, all growth and decay, is governed by the motions of the sun, moon, and planets through the zodiac.[75] Thus, though he provided no detailed astrological theory, Aristotle clearly left the way open for strong interactions between his own cosmology and astrological doctrines in Islamic and medieval thought. And these interactions probably played a significant role in reactivating interest in Aristotle during the twelfth century.[76]

SUMMARY: THE TRIUMPHS AND DEFEATS OF SCIENTISM IN CLASSICAL GREEK CULTURE

When we look at classical Athenian culture in connection with the question of the degree of integration between the attitudes and activities associated with scientific specialties and those of other crucial specialties in the culture, a remarkable story unfolds. Until the first half of the sixth century B.C., for most purposes, there simply was no set of attitudes and activities within Greek culture that could reasonably be called scientific. By the end of the fifth century scientific techniques and modes of thought had not only come into being but were assimilated by many avant-garde intellectuals to form the very basis of theological concerns (Xenophones, Democritus, Protagoras), of political concerns (Antiphon, Gorgias), and of literary expression (Euripides). This integration generated an intense conservative reaction (Aristophanes, Diopeithes) that probably represented the dominant view among Athenians who were not within the intellectual elite.

Finally, by the last half of the fourth century, scientism seems somehow triumphant among Athenian intellectuals. Each of the major intellectual schools—the Porch, the Academy, the Garden, and the Lyceum—presented its own vision of the world. These differed in myriads of ways; and in at least the Porch and the Garden, the major concerns were principally moral and ethical rather than mathematical, physical, or biological. Yet, one thing unifies all of the schools: for the central doctrines of each were created in response to problems raised and explored in connection with earlier Babylonian or pre-Socratic scientific attempts to understand the nature of the physical universe.

To the extent that we can say scientism was triumphant among Athenian intellectuals we must acknowledge that many of the conditions that brought about that triumph were totally independent of the specific character of scientific attitudes and activities. In particular, the interest in finding new explanations and understandings of social and political institutions was clearly encouraged by the failure of the earlier, and aristocratically oriented, mythological and poetically expressed ideologies to serve the needs of what we can reasonably call a rapidly emerging commercial society.

This last fact involves a curious irony that is probably also related to some of the very real and important limitations to the integration of the two longest-lived schools of Athenian thought—those of Plato and Aristotle—into broader aspects of Athenian, and more broadly speaking, Hellenistic culture. If we consider the overtly political content of Platonic and Aristotelian ideas, rather than the methods used to justify that content, we find that each produced a celebration of the small, independent, Greek polis, or city-state. But these works were produced just as the independent Greek polis effectively disappeared from the world with the rise of the Alexandrian and Roman Empires. Certain features of Platonic theory could be adapted to serve the ends of Alexandrian and Roman rulers; and aspects of Aristotle's analysis of the Athenian constitution could be adapted by Roman republicans. But in a fundamental way Platonic and Aristotelian political theory was almost as inappropriate to Hellenistic political realities as Homeric ideas had been to fifth-century Athens. In fact, the older poetic ideologies were much more appropriate to Hellenistic, Macedonian, and Roman militarism, kingship, and imperialism than those of the later Athenians; so it should cause little surprise that the Hellenistic period saw a tenacious revival of the older literature, and that Homer should have appeared to most men living in the third century B.C. as a vastly greater and more important figure than either Plato or Aristotle. There is even some reason to believe that the explicit opposition to tyranny on the part of all Athenian philosophical schools led to a self-conscious opposition to philosophers and philosophy by most Hellenistic political leaders.

There is one further feature of the aristocratic attitudes of both Plato and Aristotle that may be critical to an understanding of why the blatant scientism of the fourth-century Athenian intellectuals found itself relatively little integrated into the central features of daily religious and political life and institutions. Both Plato and Aristotle contributed centrally to the creation of a new cultural specialization—that of the learned man or philosopher who stands apart from society as an essentially disinterested and neutral observer. It is arguable that both men spoke in favor of the application of philosophical knowledge in the service of the community. But it is also the case that each had praised the disinterested search for truth above all other individual human activities. In that sense, both contributed to an attitude that has become at least one important element in the Western scientific tradition, even as we know it today; this attitude encouraged the isolation of scientific attitudes and activities from other cultural concerns almost until the Renaissance.

Prior to the fourth century Athenian emphasis on the life of the scholar as a life of dedicated contemplation, all education and learning had had a direct and immediate social goal as its primary explicit aim. The "scribal" education of priestly and governmental officials in the ancient Near East had clearly aimed at the direct integration of learning with economic, religious, and political life; and the gentlemanly education in virtue and rhetoric that flourished in classical

Athens, and continued in only slightly modified form among Hellenistic Greeks and Romans, was clearly education for citizenship. But
the philosophical and scientifically based education initiated in the
Academy, developed in the Lyceum, and given its most spectacular
expression at the Alexandrian Museum under the Lyceum-trained
Strato of Lampascus, was education in the service of pure scholarship. The magnificent intellectual achievements of such Alexandrian scholars as Archimedes and Ptolemy demonstrate the power
of such an ideal to stimulate the growth of knowledge—when it is
given adequate financial support. But the near total failure of Alexandrian science to find any immediate applications to productive
activities, and the fact that its most spectacular productions were
almost totally abandoned by the Roman and European world for
nearly a millennium, indicate what can happen when the scientific
specialty becomes almost totally divorced from the larger culture in
which it develops.

The Neoplatonist, Plutarch, tells us that Archimedes

> possessed such a lofty spirit, so profound a soul, and such a wealth of
> scientific theory, that although his inventions had won for him a name
> and fame for superhuman sagacity, he would not consent to leave
> behind him any treatise on this subject, but regarding the work of an
> engineer and every art that ministers to the needs of life as ignoble and
> vulgar, he devoted his earnest efforts only to those studies the subtlety
> and charm of which are not affected by the claims of necessity. These,
> he thought, are not to be compared with any others; in them the subject
> matter vies with the demonstration, the former supplying grandeur and
> beauty, the latter, precision and surpassing power.[77]

Even if this statement is apocryphal and misleading with regard to
Archimedes, it is symptomatic of a widespread condition that helps
to explain why Hellenistic science and scientism remained intellectually fascinating but socially unimportant phenomena.

No doubt many considerations were involved in the related
decline of Greek science and the failure of Greek culture to exploit
much of the eminently exploitable knowledge that was produced in
the Hellenistic period.[78] But among the significant factors, one must
surely acknowledge the fact that for ideological reasons, Greek
scientists seldom sought ways of developing practical consequences
from their discoveries. Nor did they seek support for their efforts on
the grounds of utilitarian considerations. To the extent that they took
seriously their Aristotelian learning, in fact, they intentionally
avoided issues related to the productive utilization of their knowl-

edge. Michael Avi-Yonah has eloquently expressed the consequences of these attitudes.

> The consequences of this "great refusal," this self-imposed limitation of idealistic philosophers, were decisive in two directions. On the one hand the blue waters of the Mediterranean remained unspoiled by the rejects of an industrial civilization. The rivers continued to flow cleanly into the sea; no smoke or gases contaminated the pure atmosphere of Greece or hid heavens from the sight of mortals. On the other hand, the Greek culture, for all its idealism and the clever talk, was forced to submit almost defencelessly to the brute force of peoples unable to ratiocinate, but capable to strike (sic) the harder.[79]

As a twentieth-century inhabitant of the Los Angeles basin, I envy the Hellenistic peoples' access to clean air and clean water; but I fear that now, as then, the consequences of any "great refusal" to integrate science with other aspects of our culture is likely to have disastrous as well as delightful consequences.

Scientism and Antiscience in Early Christianity, About A.D. 100 to 600

Scientific knowledge played some role in the development of military technology, agricultural improvement, and civil engineering during the Hellenistic period and the Roman Republic and Empire. But for most purposes, the European world simply ignored the scientific aspects of its Greek inheritance for nearly a millennium beginning around 100 B.C. Nor was there any native western European scientific tradition—though there was an important Celtic tradition of technological innovation without any apparent self-conscious theoretical basis.[1]

Only in connection with religious development was there a serious analysis and use of Greek philosophy in general, and of its scientific elements in particular. An interest in the cosmological speculations of Plato and the Stoa was a central feature of almost all significant religious movements of the late Hellenistic era; and this was no less true of Christianity than of its competitors. John Hermann Randall, Jr., has quite rightly pointed out that modern Christians tend to be ignorant of the "faith in science" that informed the early history of their doctrines.[2]

Early Christian attitudes toward all pagan learning were often ambivalent. The expression of those attitudes varied tremendously, both from person to person and within the writings of any given thinker, depending upon his rhetorical aims at the time. Yet, beginning with Justin Martyr (ca. A.D. 100–165), a converted Greek-speaking Syrian Platonist, a tradition of Christian thought grew up that was generally favorable to Hellenistic philosophy, deeply tinged with Platonic metaphysics, and forced by circumstances to focus attention on the relationship between God and the natural world; it was centered on the Christian school of Alexandria. The men associ-

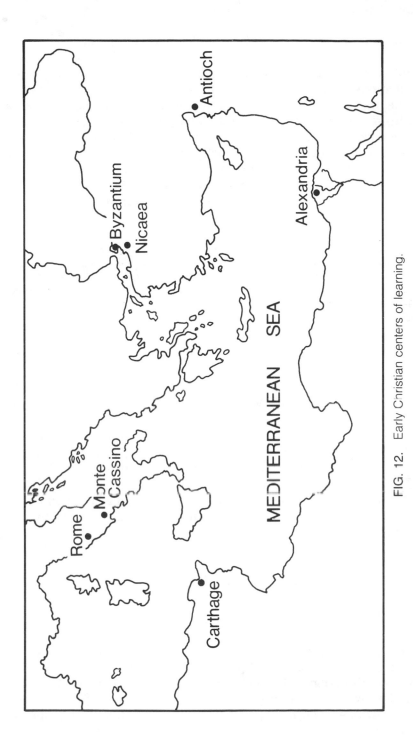

FIG. 12. Early Christian centers of learning.

ated with this school from the second to the fourth century were not scientists, though some seem to have been formally trained in medicine. Indeed, all of them warned against such an intense fascination with the natural world that religious purposes got lost. But all of them became intelligent consumers of Hellenistic science; and collectively they encouraged the Christian maintenance of a substantial amount of scientific knowledge from antiquity and the close linking of Christian doctrine with cosmological thought.

Origen (186–254) expressed the pervasive attitude of the school in a letter to one of his students.

> My desire has been that you should exercise all your natural ability in a constructive spirit, but with Christianity as your goal and aim. I wish you therefore to take over from Greek Philosophy whatever studies can be made encyclic and preparatory to Christianity and from geometry and astronomy whatever will prove useful for the interpretation of Scripture. I hope that what the sons of the philosophers say about geometry, music, literary study, rhetoric, and astronomy, that they are handmaidens of philosophy, we also may say of philosophy itself in relation to Christianity.[3]

Two special considerations insured that Christians living at Alexandria during the formative period of Christian doctrine would try to utilize pagan philosophy—especially cosmology and writings on scientific method—in explicating and interpreting Christianity. One of these was the very intimate but antagonistic relationship between Christianity and Gnosticism among the Eastern cults, extending their influence in the Greco-Roman world. The other was the special cultural and educational background of a small number of early Christians who produced written works and trained the new church's leadership.

CHRISTIAN VERSUS GNOSTIC ATTITUDES TOWARD THE PHYSICAL WORLD

To understand the growing young religion of Christianity one must be aware of the general religious revivalism at virtually all levels of society throughout the Mediterranean basin in the late Hellenistic and early Roman Empires.

At one level, highly philosophical and moral "religions" like

Stoicism and Epicureanism spread through the well-educated social elite classes. Outgrowths of the Greek philosophical schools, these religious movements lacked one key feature that was central to the more dynamic cults that were growing in the Hellenistic world—they did not emphasize "gods that listen." That is, they did not offer hope of a special helping relationship between gods and humans. Epicureanism effectively denied the relevance of the gods altogether; and Stoicism was highly skeptical of their existence. Thus the prayers of the Stoa often began with the formula, "Oh, God, if there be a God."[4] Both religions were grounded in materialistic and fatalistic philosophies, and both preached a kind of resignation to events. Evil was to be avoided by shunning excesses of all kinds and by cultivating a kind of dispassionate "apathy," which insured that no high expectations would be disappointed. Unavoidable evils were to be born with calm acceptance. In effect, the only significant threat posed to Christianity by such religions developed out of the criticisms and ridicule of Christian doctrines produced by men like the Epicurean scholar, Celsus (flourished A.D. 178), who sought to defend philosophically advanced paganism from the incursion of Eastern mystery religions like Mithraism, Judaism, and Christianity. One other highly esoteric and deeply mystical philosophical religion—Neoplatonism—emerged in parallel with Christianity, but its doctrines merged so rapidly with those of Christianity, and vice versa, that for present purposes we can ignore it as an autonomous movement.

A second religious factor crucial to Christianity was Judaism. Judaism was central to Christianity because Christ's role was grounded in the prophecies of the Jewish canonical texts—what Christians call the Old Testament—and because Christ was born a Jew. Thus Christianity was grafted onto the Jewish tradition; and Christian interpretations of scripture had to confront a tradition of Jewish exigesis (explanation of the text) and scholarship.

The challenges posed by Judaism to Christianity were particularly critical at Alexandria, where Philo Judaeus (flourished A.D. 40) had developed an impressive Platonizing interpretation of the Old Testament writings; where the Jewish writings had been translated from Aramaic and Hebrew into Greek; and where pagan scholars and Gnostics had developed a strong school of criticism of the anthropomorphic tendencies of Judaism, and the severity and exclusiveness of the Jewish God.

Though Judaism was of central importance to Christianity because it provided much of the substance of Christian doctrine, and some of its severest liabilities from the point of view of pagan scholars, we do not discuss it here because it had little immediate bearing on Christian attitudes toward science. Instead we focus on the third major religious movement of the first four centuries—a movement that challenged Christianity most intensely because it provided the most direct competition for men's attention and for men's souls.

To an outsider, Christianity must have appeared initially as only one minor variant among many Eastern cults that shared common features. All were henotheistic, in that they preached the existence of one supreme, good God—though Christianity became vehement in denying the independent existence of lesser, evil gods. All were universal, in the sense that they sought converts from all classes, races, and nationalities. All were religions that promised salvation in an afterlife. Almost all preached that this salvation came through a mediator who was at once somehow both human and divine. All stressed the need to atone for sins. All were grounded in sacred texts, presumably revealed by God to men. And almost all offered the central rituals of baptism and the Eucharist (a holy supper in which the symbolic partaking of the flesh and blood of the Savior plays a role). In some cases (e.g., in that of Valentinian Gnosticism, which rose at Alexandria in the second century) many of the doctrines and rituals of the other cults were directly borrowed from Christianity. But in most cases, they were associated with much earlier Persian religions.

The Gnostic cults were far and away the most important of these Eastern religions in terms of their significance for Christianity because they were most directly related to Christian thought, and because their doctrines were ably taught by two great preachers, Valentinus and Basilides, at Alexandria while Christian doctrine was being consolidated during the second and third centuries. In terms of stimulating Christian interest in science, and in the *Timaeus* in particular, Gnosticism was all important. Early Gnosticism (see chap. 2) was grounded in the belief that the material world is both deterministic and fundamentally evil. R. B. Tollinton summarizes the Gnostic attitude toward matter as follows:

> Matter is contrasted . . . with form. It is the lowest constituent. It limits the spirit. It beguiles the soul that enters into it. It drags its spiritual visitor down to the lower levels and even sinks it in the mud. Constantly the epithets applied to it are such terms as lifeless, shapeless, corruptible, base, despicable, shifting, defective. It is classed with vice and corruption. At best it is devoid in its own nature of all good. More often

it is the positive cause of evil, restricting ideals, thwarting the purpose of the artificer, occasioning sin . . .[5]

The Gnosticism taught at Alexandria appealed to the second part of Plato's *Timaeus* (see chap. 4) to claim that the inherent evil in matter had limited God's power to produce a good universe, and had forced him to create a world that was a place of mere trial to be transcended through the process of spiritual salvation. Alternatively, Valentinus identified the Platonic Demiurge with the Jewish Jehovah and saw him as a fallen soul who created the evil physical world out of spite, and in opposition to God.[6] The book of Genesis, on the other hand, describes the creation of a material universe characterized by God, himself, as "very good,"[7] one in which, as the nineteenth psalm assures us, "The heavens declare the glory of God and the firmament showeth his handiwork."[8] Moreover, Christians could not acknowledge that evil is imposed upon the world by matter in opposition to God's will, for God is both good and all powerful. Nor can evil come from a positive source independent of God, for there is no power in the universe other than that granted by God.

In opposition to the Gnostics, Clement of Alexandria (ca. 150–215) and Origen presented an interpretation of Genesis that interweaves scriptural and Platonic elements with scattered borrowings from the Stoa and even from their Gnostic enemies; but the Christian version of creation focuses on the basic goodness of the world. Even before the creation of the physical universe, God created the perfect immaterial universe, forming in the Son, or Logos, the ideas that constitute the model that the Platonic Demiurge had worked from.[9] Next, the Son (Logos), as active agent, played the Demiurge's role in creating the material universe into which God sowed "intelligences." Like Plato, Origen insists that, "the whole world ought to be regarded as some huge and immense animal, which is held together by the power and reason of God, as by one World Soul."[10] Moreover, Origen insists, as had Plato, that the course of natural events occurs neither by chance, as the Epicureans supposed, nor through some purely materialistic "nature," as it did for the Stoa, but through natural laws established in the beginning by God. One modern interpreter of Origen puts the issue this way:

> Laws regulating the change of bodies were established by God, and it is therefore in accordance with an original divine *fiat* of this kind that [apparently] spontaneous changes take place, such as physiological changes in the human body at puberty, the movements of animals,

plants, fire, and water, and the incitement of activity which impels a
spider to weave a web.[11]

The world is not strictly deterministic, for God could will the
suspension of his laws at any time and for any purpose; but its
ordinary concourse is knowable and predictable.

As the creation of an infinitely good God through the agency of
his Son, the world could not but be good. Then how did the Alexandrine Christians explain the apparent evil, which was the focus of
Gnostic understanding of the world?

When God created the intelligences, he chose to make them
free. All were created initially equal; but through an unexplained
process, which is suspiciously like that of Plato's metempsychosis,
some intelligences sinned and fell into lower states while others
became even better. The worst of the first group became devils,
fiends, and archfiends; whereas the best of the latter group became
stars, angels, and archangels. But those in the middle, and those
whose sins were minor and redeemable, became the souls of living
beings capable of salvation, but only through a form of discipline in
this world.

In order to provide the proper corrective treatment for each
stage of fall from grace and the proper encouragement for each stage
of improvement, the world was created "infinitely various—like a
great house in which are vessels of Gold and Silver, Wood and Clay,
some to honor and some to dishonor. . . . Wherefore neither will the
Creator seem unjust when he distributes to each his earthly lot."[12]

We cannot know in precise detail the reasons for every apparent
evil. As Origen tells us,

> There are things in the creation hard to understand, or even undiscover-
> able for human beings. We are not in consequence to condemn the
> creator of the universe just because we cannot discover the reasons for
> the creation of Scorpions and other venomous beasts.[13]

It is enough for the good Christian to recognize that the apparent
evils of the natural world are in fact aspects of an ultimately good
design to allow for the salvation of souls that choose freely the real
evil of sin. Even this realization, of course, requires some study and
effort on the Christian's part.

As a result of the conflict between Christian and Gnostic over
the nature of the physical world and the origin of evil, Christian

teaching was forced to focus attention on the nature of the cosmos and God's relation to it. And it was forced to emphasize the good aspects of the physical universe in spite of the fact that Christianity itself sought ultimate transcendence of the physical world. Some Christian opponents of Gnosticism were, in fact, driven to such an upgrading of the physical world that they insisted literally that heaven would come *on earth*; but such men, including Justin Martyr, were in a distinct minority.

It would be wrong to see early Christian concern with the physical world as an overriding concern. After all, the central doctrines of Christianity deal with the salvation of the soul. Even such a cosmos-conscious Christian as Origen slipped occasionally into speaking of matter as "the Devil's mistress" and into writing that, "the soul has descended. Let it shun the World and all its evil and keep itself uncontaminated as may be until the time comes for its return."[14] On the other hand, interest in the physical universe was more than a short-lived and superficial concern. According to all of the major early Greek fathers, there are but two ways of knowing God. One is through his revelation to man through the Word; the other is through the evidence of his activities manifested in the creation. For some, the Word clearly took precedence over the world as a source of valuable knowledge. But this was not necessarily true of the Alexandrine fathers who had been pushed by their Gnostic enemies to a particularly intense consideration of the latter.

In *On First Principles*, his most concerted attack upon Gnosticism, Origen wrote:

> If we see some admirable work of human art, we are at once eager to investigate the nature, the manner, and the end of its production; and the contemplation of the works of God stirs us with an incomparably greater longing to learn the principles, the method, and the purpose of creation. . . . This desire, this passion, has without doubt been implanted in us by God. And as the eye seeks light, as our body craves food, so our mind is impressed with the characteristic and natural desire of knowing the truth of God and the causes of what we observe.[15]

Moreover, our naturally sought knowledge of the world leads us toward a knowledge of God, for as Plato had suggested, the divine wisdom, "from actual things and copies, teaches us things unseen by means of those that are seen, and carries us over from earthly things to heavenly."[16]

For Origen, in fact, the phenomena of nature are neither more nor less important and clear as a source of our understanding of the divine than the scriptural revelation.

> We know that in its work of creation the divine skill is displayed not only in the heavens and the sun and moon and stars, as it pervades the whole of their mass, but also on the earth it operates in the same way in any common substance, so that the bodies of the tiniest animals have not been neglected by the Creator. . . .
>
> It is so in the plants and the soil; in each is an element of design affecting its rocks and leaves and the fruit it can bear, and the characteristics of its qualities. Now we take the same view of all scriptures that proceed from the inspiration of the Holy Spirit . . . *He who has once accepted these scriptures as the work of Him who created the world must be convinced that whatever difficulties in regard to creation confront those who strive to understand its system, will occur also in regard to the scriptures.*[17]

Theophilus, patriarch of Alexandria from A.D. 385 to 412, carried the notion that the essence of Christianity was to be found in an understanding of the natural world to its extreme, writing that, "All Christianity is to be found in the first six chapters of Genesis,"[18] (i.e., in the story of the creation). While few Christians prior to the seventeenth century placed quite such a heavy burden upon natural phenomena as evidence for Christianity, almost every major theologian of the ancient and medieval world did so to some degree. Most frequently—at least after the writings of St. Basil (A.D. 330–379)—their comments on cosmology and Christianity were offered in Hexamera, sets of commentaries on the six days of creation as they appear in Genesis. No form of Christian literature is more extensive and continuous than that of the Hexamera. Various controversies have waxed and waned, and emphases have shifted over time. But from Basil through the Latin fathers, including St. Ambrose and St. Augustine; through the works of Isadore of Seville, Walafrid Strabo and the great Englishman Bede in the early Middle Ages; through the writings of twelfth-century scholars like Thierry of Chartres and Robert Grosseteste; and well into the fourteenth century in works like Henry of Langenstein's *Lecture super Genesim*,[19] the book of Genesis both demanded Christian consideration of the nature of the physical world and provided an opportunity for Christians with scientific leanings to explore their interests.

THE "CHRISTIAN GNOSTIC" AND GREEK SCIENCE

If the conflict between Christian and Gnostic stimulated an interest in nature and gave rise to a tradition of commentaries on the creation, the particular scholarly and scientific form that the interest in nature took was determined in large part by the social and educational background of the Alexandrine fathers and their intellectual descendants.

During the early centuries of Christianity, Alexandria was the intellectual center of the Greco-Roman world. It was the site of the great museum, of the greatest Jewish scholarly community, and of the Christian catechetical school. Moreover, although Christianity may have been to some degree a religion that drew primarily from the illiterate masses (this is debatable), virtually all of the fathers who left a significant mark on doctrinal development—whether from Alexandria, Syria, Carthage or Rome—had excellent scholarly training in pagan literature and science. Justin Martyr, for example, had been a teacher of Platonism before his conversion to Christianity. Clement of Alexandria studied philosophy at the Christian School of Alexandria. Origen studied under both Clement and Ammonius Saccas, the founder of Neoplatonism; and we know from his students' letters that Origen taught geometry, physiology, and astronomy as well as summaries of other philosophical systems "except for that of the Godless Epicurus." Tertullian (ca. 160–230) was a student of literature, philosophy, and law before his conversion; Lactantius (from ca. 313) taught rhetoric; St. Basil (ca. 330–379) studied medicine; and St. Augustine (354–430) taught rhetoric and became a convert to Neoplatonism before he became a Christian at age thirty-three.

These men were often—though seldom admittedly—proud of their learning and committed to some degree to the Platonic and Aristotelian notions of the dignity and value of contemplation. They were also very sensitive to pagan claims that Christianity was a religion for ignorant illiterates. In response to such claims Clement and Origen developed a theory that distinguished levels of Christianity, based on whether Christian doctrines were accepted on blind faith or on a reasoned knowledge of their full meaning. Both men saw faith as proper to the initial and lower stages of Christian life. He who lived in faith alone, lived a life of fear and hope. Such a life could be one of holiness and salvation, but it could not be one of joy and righteousness. For the Christian scholar, however—the true "Christian Gnostic"—there was a higher life of knowledge, love, and righteousness.[20]

How does one attain the status of the highest Christian life? One must be pure in heart, of course, and one must study the scriptures; but one must also study the Greek philosophers and scientists. Clement of Alexandria points to this latter need most clearly.

> Some people, thinking they are gifted, deign neither to touch philosophy or dialectic nor to learn physical science. They require bare faith alone. Just as if they expected, after giving no attention to the vine, to pick bunches of grapes immediately from the beginning. The Lord is allegorically called the vine, from which, only with care and logical husbandry, the fruit is to be gathered.[21]

Elsewhere Clement praises astronomy in particular because "this science makes the soul particularly quick to understand, of clear insight into truth, and capable of refuting falsehood. It enables the soul to find agreements and relations so as to hunt out likeness in unlike things."[22]

In general, Clement argues, "philosophy is not a gift of the devils, but of God through the Logos, whose light ever beams upon his earthly image, the intelligence of man. [It is] like the burning glass whose power of kindling is borrowed from the sun."[23] Here is a tremendously fruitful and important set of classical metaphors combined into one Christian doctrine. Knowledge, which not only leads to, but even constitutes, man's supreme happiness, is ultimately the gift of God through the "Logos," a word which simultaneously means the reason of the philosophers, the scriptural Word, and the highest manifestations of Christ.[24] For Clement and Origen, as it was for Aristotle, knowledge is particularly important precisely because it is in his intellect that man most closely images the divine. Finally, in Clement's passage the relation between man's intellect and God's is illuminated by the Platonic sun-vision analogy of the *Republic*.

One effect of the Alexandrine Christian's interest in Greek philosophy was to ensure that the knowledge of nature they appealed to in their fight with Gnostics and other pagans, and in their exigetical commentaries on scripture, would come much more often from the Platonic, Stoic, and Aristotelian traditions than from any direct investigations of their own.

A second effect of their learned approach to Christianity was to undermine any attempt to interpret scripture too literally. Scripture was often superficially crude or absurd. "Nothing," according to Clement, "is to be believed which is *unworthy* of God."[25] In the face of the literal impossibility of certain scriptural passages, those passages had to be assumed to be carriers of hidden allegorical meaning, transparent only to a learned elite. Through the use of allegorism, the Alexandrine fathers could develop techniques for reconciling scripture and scientific knowledge, and they could extend the range of applicability of all of their learning. Though this allegorism was held suspect by some powerful church intellects, it was adopted by Augustine and turned into a fundamental characteristic of medieval scriptural interpretation. This is not the least of the characteristics of Christian thought attributable to early interaction with a scientific attitude.

A third, and perhaps the most crucial effect of the Alexandrine Christian's interest in secular philosophy is that techniques of argumentation, analytic categories, and examples from Greek mathematics and science provided the very structure, language, and basic metaphors for early Christian theological discourse.

Nowhere can the impact of scientific patterns of thought on theological issues be seen better than in the writings of Clement of Alexandria. In his attacks upon pagan polytheism, Clement sounds precisely like the Aristophanic Socrates (see chap. 3) when he attacks the deification of idols and natural phenomena.

> Surely it is plain to every one that they [i.e., idols] are stones just as Hermes himself. And as the halo is not a God, nor the rainbow either, but conditions of the atmosphere and clouds, and precisely as a day is not a god, or month, nor year, nor time which is made up of these; so also is neither sun nor moon, by which each of the before-mentioned periods is marked off. Who, then in his right mind would imagine such things . . . to be Gods.[26]

Clement borrowed even more from the scientific tradition in creating positive Christian doctrine than in his attacks on pagans. To illustrate these borrowings we briefly consider just two among many issues—the problem of understanding God's character, and the problem of justifying the crucial role of faith as a starting point for all Christian knowledge.

Clement begins his discussion of God by misquoting the *Timaeus* to the effect that it is not only hard to discover the Creator, but, "having found him it is impossible [rather than difficult] to speak about him."[27] This claim would seem to leave little to say; but Clement plunges on by claiming that we can discover God by analyzing what he is not. To explain how this approach can work he uses a version of Aristotle's abstractive approach to geometry.

> We take away from physical body its natural qualities, stripping it of the dimension of depth then of breadth, and after these, that of length. For the point is left in unity, as it were, with position, and if we remove position from it, unity is perceived . . . if we do these things we shall reach in some way the perception of the Almighty, knowing not what he is, but what he is not. But shape and movement, or standing still, or throne of place, or right hand or left hand, are in no way whatever thought of as belonging to the Father of all things, even if those things are written. . . . *The first cause*, then is not in any place, but above place, and time, and name, and thought.[28]

In this way Clement throws off the anthromorphism of Judaism and quietly identifies the Christian God at once with a kind of Anaximandrian placeless amorphous unity and with Aristotle's timeless and motionless First Cause. Ultimately, of course, God cannot be understood through scientific demonstration, "for," as Clement says, following Aristotle, "this depends on prior and more readily known principles, and there is nothing prior to the unoriginated."[29] But we *can* clear ourselves of false ideas in this manner and prepare ourselves to receive knowledge grounded in faith and granted through God's grace.

Clement's defense of faith as the starting point for Christian Gnosis is presented in the form of seven arguments, of which the three most important derive from Plato's and Aristotle's theories of scientific method. In the first argument Clement cleverly uses the implications of Aristotle's theory of science to point out that *all* knowledge ultimately rests in faith of some kind.

> Either all things are in need of demonstration, or some are believed of themselves. But if the former is the case, and we ask the demonstration of each demonstration, we shall go on ad infinitum and demonstration will be brought to nothing. But if the latter is the case, those very things which are believed of themselves will become first principles of demonstrations. Indeed the philosophers all agree that the first principles of all things are indemonstrable. So that if indeed there is demonstration, there is an absolute necessity that there is something prior which is believed from itself and is called first and indemonstrable. Consequently, all demonstration is referred back to indemonstrable faith.[30]

The second argument begins where the first left off, and appeals to Plato's theory of knowledge through a syllogistic argument. If Plato's *Meno* is correct, all knowledge is founded on what was previously known. But first principles are not grounded in what was previously known (Clement purports to offer an inductive proof of this claim). Therefore, first principles cannot be knowledge, and, "the first principle of the universe can be reached by faith alone."[31]

This argument would, of course, not hold if Aristotle's account of the induction of first principles from sensory experience were true. But Clement is far too much of a Platonist to accept Aristotle's account.

> [Some] try to drag all things down to earth out of heaven and the unseen, literally grasping rocks and trees in their hands, for they lay hold of all things and vigorously assert that that alone exists which can be handled and touched. They define body and reality as the same thing. Those who dispute with them very properly defend their position somewhere above from the unseen world, forcefully declaring that certain intelligible and bodiless forms are the true reality.[32]

The third argument, which is claimed for Epicurus, expands on the first and is closely related to the mathematical method of hypothesis testing that Plato used in his works (see chap. 4).

> [Epicurus] assumes faith to be preconception of the intellect. . . . No one can investigate or raise a question, or hold an opinion, or argue without a preconception. How can one, not having a preconception of what he is after, learn about what he investigates? When he has learned he has now turned his preconception into a comprehension.[33]

The point here is that blind groping after knowledge can lead nowhere. One must make certain assumptions and follow out their consequences to discover whether they conform to reality as we understand it. The initial assumptions, or preconceptions, or hypotheses, must, according to Clement, come by faith. Thus, the Christian's claim for the necessity of faith turns out not to be an ignorant avoidance of the difficult process of seeking rational truth, as critics like Celsus maintained. Faith is rather a necessary precondition of that search as implied by the methodological writings of all major philosophical schools. Christian Gnosticism thus leads to an understanding of religious issues, which depend for both its vocabulary and its methods of argumentation on the scientific or scientistic traditions of Plato, Aristotle, and Epicurus.

The notion that Christian Gnosticism is somehow a higher form of Christianity than that attainable by the common lay person, and that ordinary Christians therefore need the guidance of a learned elite to interpret the scriptures, has always had many detractors; but the doctrines set forth by Clement and Origen were certainly effective in establishing a powerful tradition of study and contemplation, especially in Eastern (but even in Latin) monasticism; moreover, their doctrines were central in establishing the Roman Catholic Church's institutional insistence that the hierarchy of the church must be charged with interpreting doctrinal and scriptural points, which are beyond the means of lay persons to understand.

EARLY CHURCH ANTISCIENCE

In general, the early Latin fathers were much less sympathetic to pagan science and philosophy than their Alexandrian counterparts. To some extent, their antagonism to Hellenistic learning reflected a broad Roman antiintellectualism. But the churchmen had special reasons for opposing pagan science and philosophy. They quite correctly saw heresies like Gnosticism as grounded in astronomy and in the doctrines of Plato and Aristotle. Thus, St. Irenaeus bitterly comments, "They [the heretics] strive to transfer to the treatment of matters of faith that hairsplitting and subtle mode of handling questions which is, in fact, a copying of Aristotle."[34] And Tertullian asks, in a famous rhetorical passage, "What has Athens to do with Jerusalem?" to which his obvious reply is, "Nothing whatever."[35]

To most of the Latins it seemed all too easy for a scholar to be seduced away from the primary aims of Christianity—those associated with salvation—into a pursuit of natural knowledge that was not simply pointless from their perspective, but encouraged the sins of pride and arrogance. St. Ambrose, for example, asks,

> What shows such darkness as to discuss subjects connected with geometry and astronomy, to measure the depths of space, to shut up heaven and earth within the limits of fixed numbers, to leave aside the grounds of salvation and to seek for error.[36]

Arnobius reiterates:

> What is it to you . . . to examine, to investigate who made man, what origin souls have, who conceived the causes of evils, whether the sun is

larger than the earth, or measures a foot across, whether the moon shines by the light of another or by its own beams? . . . There is no gain in knowing these things nor any loss in not knowing them. Leave these things to God and allow him to know what, wherefore, and whence something is . . .[37]

But Tertullian has the last laugh in retelling one of the great absent-minded professor stories.

> It . . . served Thales of Miletus quite right when, stargazing as he walked with all the eyes he had, he had the mortification of falling in a well, and was unmercifully twitted by an Egyptian who said to him, "Is it because you found nothing on earth to look at, that you think you ought to confine your gaze to the sky?!" His fall, therefore, is a figurative picture of the philosophers, of those, I mean, who persist in applying their studies to a vain purpose, since they indulge in stupid curiosity on natural objects.[38]

Perhaps even more crucial than their concern for the pointlessness of natural philosophy was the Latin fathers' special desire to establish Christian faith not as the mere starting point for Christian belief, but as totally sufficient. Much more than the Alexandrines, the Latin fathers faced a relatively poorly educated population of Christians. Their worries were often those of Aristophanes in the *Clouds*. Arguments and interpretations of the causes of natural phenomena seemed to lead most often, not to a strengthening of Christian commitment, but to an erosion of faith. Thus, says Irenaeus, "If we leave some questions in the hands of God we shall both preserve our faith uninjured and continue without danger."[39] And Lactantius repeats: "If you believe, why then do you require a reason, which may have the effect of causing you not to believe?"[40]

The real problem is that for Christians in the second and third centuries A.D., as for Greek polytheists in the sixth and fifth centuries B.C., attempts at scientific *understanding* too often seemed to lead to religious *undermining*.

The letters of Paul the Apostle had offered the grounds for a particularly antiintellectual and faith-oriented Christianity. The Latin fathers chose to ignore Romans 1:14, where Paul writes, "I am debtor both to the Greeks and to the Barbarians," and instead focused on 1:17, "The just shall live by faith." Irenaeus writes that "It is therefore better and more profitable to belong to the simple and unlettered class and by means of love to attain nearness to God."[41] But once again the most spectacular presentation of the Latin doctrine of the supremacy of faith comes from the great rhetorician Tertullian.

> And the Son of God died; it is by all means to be believed, because it is absurd. And he was buried and rose again; the fact is certain because it is impossible.[42]

No reason could conceivably establish the death of an immortal or the rebirth of a dead mortal. Thus, the death and resurrection of Christ become the touchstone of a need for faith—a faith which stands opposed to, rather than as a foundation for, reason.

THE AUGUSTINIAN RESOLUTION

The tension and sometimes overt conflict between reason and faith, Alexandrine and Latin, intellectual elite and antiintellectual common man, which was reflected in, and to some extent generated out of, early Christian responses to classical scientific thought, has been a recurrent feature of Western culture—one replayed in particularly virulent forms in both the thirteenth and late seventeenth centuries. But in its first Christian version the conflict was temporarily resolved by the great Carthaginian, St. Augustine, whose works became singularly important for the Latin medieval church. Augustine was undoubtedly inconsistent in his attitudes toward pagan learning in general, and pagan science in particular. Yet the overall effect of his inconstancy was to produce an uneasy but relatively stable tension between Christian emphases on faith and learning.

Augustine had been a Manichean Gnostic during his youth and had passed through a period of intense Neoplatonism before he converted to Christianity in A.D. 387, at the age of thirty-three. He was well versed in classical literature and philosophy, but he shared his Latin colleagues' emphasis on faith and salvation.

On the one hand, Augustine shared the Latin sense that scientific knowledge was of limited value and certainty.

> Nor need we be in alarm lest the Christian should be ignorant of the force and number of the elements—the motion and order, and eclipses of the heavenly bodies . . . and a thousand other things which those philosophers have found out, or think they have found out—for even these men . . . have not found out all things; and their boasted discoveries are oftener mere guesses than certain knowledge. It is enough for

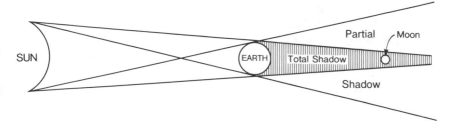

FIG. 13. Diagram illustrating the explanation of lunar eclipses.

the Christian to believe that the only cause of created things is the goodness of the Creator, the one true God.[43]

Astronomy in particular received harsh treatment because of its association with the dangers of physical determinism:

> This knowledge, although in itself it involves no superstition, renders very little, indeed almost no assistance in the interpretation of Holy Scripture, and by engaging the attention unprofitably is a hindrance rather; and as it is closely related to the very pernicious error of the diviners of the fates, it is more convenient and becoming to neglect it.[44]

On the other hand, Augustine believed that some learning, including scientific learning, is desirable to aid in explaining the scriptures, and to aid in defending Christian doctrines against attack. Works on geography, animals, trees, herbs, stones, and other bodies may all be useful in this sense.[45] Geometry is valuable because "this type of discipline trains the mind to discern more subtle matters . . . and, unless I am mistaken, brings forward arguments that are very rare."[46] But, above all, the science of reasoning developed by Plato and Aristotle is useful, for, "the science of reasoning is of very great service in searching into and unravelling all sorts of questions that come up in Scripture."[47]

Augustine occasionally even admits that a certain limited amount of learning for its own sake might be valuable to the soul, "For, just as a body denied food is often ridden with sickness and scabs, maladies which indicate hunger, so the soul of the uneducated also are filled with diseases whereby they betray their malnutrition."[48]

All learning is best applied to Christian ends, and in all studies our guiding principle should be, "Not too much of anything, especially in the case of those, which pertaining as they do to the senses, are subject to the relations of space and time."[49]

It would be hard to claim that St. Augustine embraced many scientific attitudes or that he engaged in any scientific activities. Yet, limited as it was, his authorization of some training in secular studies—especially logic and mathematics—encouraged the development of an institutional format for Christian education that contained a substantial scientific component as a background to theological studies. (At one point Augustine projected writing a work on the seven liberal arts, but he finished only a small segment.)[50] Thus, logic, rhetoric, grammar, music, arithmetic, geometry, and astronomy became at least nominal prerequisites for Latin clerics in high positions. During the High Middle Ages, the institutional feature would prove to be extremely important.

The Seven Liberal Arts		The Mechanic Arts
The Trivium	The Quadrivium	
Logic (dialectic) Rhetoric Grammar	Music Arithmetic Geometry Astronomy	Medicine Architecture

FIG. 14.   Seven liberal arts and two mechanic arts according to M. Terentius Varro (c. 30 B.C.). The seven liberal arts formed the basis for the curricula in Christian educational institutions through the Renaissance.

Finally, Augustine acknowledged that study of the works of creation could lead one to contemplation of the immortal; and this leads us back to the Hexameron tradition; for it was from the Hexamera of Basil and Ambrose (Basil's almost parrotlike follower) that many of Augustine's comments on nature derived.[51]

CHRISTIAN COSMOLOGY: THE HEXAMERON OF ST. BASIL

Early Christian writings on nature were definitely addressed to theological ends and to a relatively untrained audience. So they were seldom, if ever, involved in adding to the stock of what we should call scientific knowledge. Yet, the authors of the Hexamera did succeed in producing a uniquely Christian cosmology from the amalgam of pagan and sacred literature, which they sought to synthesize. Their synthesis involved a shrewd and often critical assessment of specific pagan doctrines; and the Hexameral literature inculcated in its hearers and readers a set of commitments to scientific attitudes.

Virtually all of the important characteristics of the Hexameral tradition can be illuminated by considering the *Exigetic Homilies* of St. Basil, which were simultaneously the most popular, and among the most intellectually coherent, of early Christian cosmological treatises. Furthermore it was often through Basil's works alone that subsequent Christians saw the background to Christian cosmology.[52]

Basil thought the primary explicit reason for considering the natural world was to derive from it a knowledge of God and an appreciation for his wisdom, power, and beneficence.

> If you see the heavens . . . and the order in them, they are a guide to faith, for through themselves they show the craftsman; and if you see the orderly arrangement about the earth, again through these things also your faith in God is increased . . . Even if you consider the stone, it also possesses a certain proof of the power of its maker; likewise, if you consider the ant or the gnat or the bee. Frequently in the smallest objects the wisdom of the Creator shines forth.[53]

Elsewhere he reiterates the words of both Plato and St. Paul, writing that the natural world,

> is truly a training place for rational souls and a school for attaining the knowledge of God, because through visible and perceptible objects it provides guidance to the mind for the contemplation of the invisible, as the Apostle says: "Since the creation of the world, his invisible attributes are clearly seen . . . being understood through the things that are made."[54]

Just as the world provides a training ground for our understanding, so too does the account of its creation in Genesis, for as Basil says, "the narrative made omissions to accustom our minds to a ready understanding and to permit the rest to be deduced from slight resources."[55]

Basil demonstrates a secondary, but still important, motive for studying nature that is related to the older Alexandrine allegorizing tradition. Basil was not himself an allegorist. He repeatedly insisted that we must not be "ashamed of the gospel"[56] and that when we read "water" or "grass" we must understand those words in their common meaning rather than as symbols of some hidden spiritual entities. Yet, Basil wanted to insist that we can learn moral and spiritual lessons from the study of natural phenomena. Precisely because we stand distinguished from the lower organisms by our ability to reason, we can recognize in their behavior a kind of natural law imposed by God which we would do well to acknowledge and accept. Discussing the migration and spawning habits of fishes, Basil makes this point with great eloquence and clarity.

> You see that the divine plan fulfills all things and extends even to the smallest. A fish does not oppose the law of God, but we men do not endure the precepts of salvation. Do not despise the fish because they are absolutely unable to speak or to reason, but fear lest you may be even more unreasonable than they by resisting the command of the Creator. Listen to the fish, who through their actions all but utter this word: "We set out on this long journey for the perpetuation of our species." They do not have reason of their own, but they have the law of nature strongly established and showing what must be done . . . I have seen these wonders myself and I have admired the wisdom of God in all things. If the unreasoning animals are able to contrive and look out for their own preservation, and if a fish knows what it should choose and what it should avoid, what shall we say who have been honored with reason, taught by the law, encouraged by the promises, made wise by the Spirit, and who have then handled our own affairs more unreasonably than the fish? . . . Let no one allege ignorance. Natural reason, which teaches us an attraction for the good and an aversion for the harmful is implanted in us. *I do not reject examples drawn from the sea, since these lie before us for examination.*[57]

Likewise, Basil saw in the behavior of the plants of the field, the creatures of the land, and the birds of the air illustrations of moral imperatives for human life. This moral use of nature was a subdominant theme for Basil, but it was the foundation of an intense Christian interest in natural history, which manifested itself in the near contemporary *Physiologus* literature.[58]

The third major motivation for Basil's treatment of cosmology was clearly apologetic; it was intended to defend Christian dogma against potential antagonists—Gnostics, Astrologers, Platonists, Aristotelians, Stoa, and Epicureans alike. This motive is implicit in the very structure of Basil's *Exigetic Homilies* on Genesis; virtually every comment can be seen as a critique of some specific pagan doctrine or a shoring up of the Genesis account against charges of inconsistency or implausibility.

In many ways Basil's Hexameron, and those that followed it in the Christian tradition, can best be understood as cosmological treatises much like the *Timaeus* of Plato or the *De Rerum Natura* of the Epicurean, Lucretius. They are neither more nor less eclectic and syncretistic than the pagan treatises; they are little, if any, more moralistic in tone than Lucretius' work; and they are probably not significantly more dependent on metaphysical or other philosophical or religious presuppositions than the pagan works, although they are probably more open and explicit about them. It is true that the early Hexamera show less concern with technical details than most

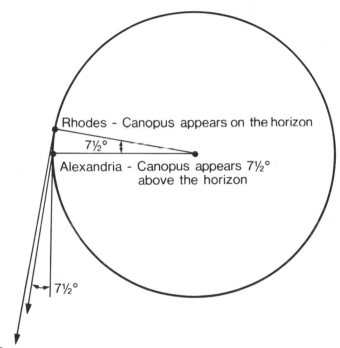

FIG. 15. Posidonius's method of calculating the circumference of the earth cited by St. Basil. It was known that Alexandria and Rhodes are on the same great circle through the poles and that they are 3,750 stades distant from one another. From the figure it is evident that:

$$\frac{\text{Earth's circumference}}{3{,}750 \text{ stades}} = \frac{360°}{7.5°} \text{ , or,}$$

The Earth's circumference = 180,000 stades.

pagan treatises. Thus in commenting about the relation between Genesis and pagan accounts of the earth, Basil says,

> I shall not be persuaded to say that our version of the creation is of less value because the servant of God, Moses, gave no discussion concerning the shape and did not say that its circumference contains one hundred and eighty thousand stades, nor measured how far its shadow spreads in the air when the sun passes under the earth, nor explained how, when this shadow approaches the moon, it causes the eclipses.[59]

Basil's account, of course, shows that *he* is aware of these details even if they were not contained in Genesis; and later medieval Hexamera tend to fill out details of the world picture, little by little.

For all of their similarities with other cosmological treatises, the Hexamera had one crucial characteristic that often makes them fundamentally unscientific in spite of their tremendous dependence on, and incorporation of, scientific knowledge and methods.

All early Hexamera are at least as centrally explanations and defenses of the *biblical account* of the creation as they are interpretations of the phenomena of the world itself. They do not ignore phenomena—in fact much ingenuity is devoted to demonstrating that phenomena conform to what may initially seem to be absurd or inconsistent biblical statements—but the *focus* of concern is often the text of Genesis; and when there are apparent conflicts between Genesis and what it is intended to account for, the Christian scholars were bound to place the authority of scripture before that of their senses or their reason. On this issue Basil is both succinct and characteristic. The Bible is the revealed word of God, so "even if we do not attain to the profound thoughts of the author because of the weakness of our intellect, nevertheless, having regard for the authority of the speaker, we shall be led spontaneously to agree with his utterances."[60] Similarly, "Let us hear, therefore, the words of truth expressed, not in the persuasive language of human wisdom, but in the teachings of the Spirit."[61]

On this score, the Hexameral literature must be seen as highly qualified in its acceptance of scientific attitudes; and this qualification was of major significance in giving shape to what passed for science in the Christian West, at least into the period of the Renaissance. The emphasis both on interpreting texts and in accepting textual authority as on a par with, or superior to, the authority of the senses or reason shaped medieval scholarship, even when it became much more concerned with secular knowledge in the late Middle

Ages. It is at least in part for this reason, for example, that the vast bulk of thirteenth-century science appears in the form of commentaries on Aristotle's scientific writings or commentaries on Genesis.

So much for the motives for, and forms of, early Christian science. What was the content of the new Christian cosmology? How was it related to earlier scientific literature? And what major issues did it pose for subsequent European intellectual life? Again, Basil sets the tone for nearly a thousand years of Christian literature.

The first lecture of Basil's Hexameron is entirely devoted to a commentary on the first sentence of Genesis, "In the beginning God created the heavens and the earth." This statement provides the opportunity to review and critically assess the four major cosmological premises of Greek antiquity; those of the atomists, the Stoa, the Platonists, and the Aristotelians.

The atomists, of course, failed to acknowledge that there was a God who was the rational cause of the universe, so they blundered into imagining,

> that the nature of visible things consisted of atoms and indivisible particles or molecules and interstices; indeed, that, as the indivisible particles now united with each other and now separated, there were produced generations and deteriorations; and that the stronger union of the atoms of the more durable bodies was the cause of their permanence. . . . They were deceived by the godlessness present within them into thinking that the universe was without guide and without rule, as if borne around by chance.[62]

Just as the first sentence of Genesis contravenes all materialist theories of creation, it also directly contradicts the Aristotelian account, for it stipulates a beginning of the physical universe and of time.

> Do not imagine, then, . . . that the things which you see are without a beginning, and do not think, because the bodies moving in the heavens travel around in a circle and because the beginning of the circle is not easily discerned by our ordinary means of perception, that the nature of bodies moving in a circle is without a beginning.[63]

Similarly, the first sentence belies the Stoic notion that the Heavens *are* God and have existed from eternity.[64]

The relationship that Basil seeks to establish between Genesis and the Platonic account of creation is complex. As one might guess, Basil cannot admit that God fashioned the universe out of a preexistent matter whose characteristics limit God's options. In his words, God cannot be "unsuccessful and inefficacious because of the deficiency in matter."[65] Thus, Basil insists that the Bible used the word "created" rather than "fashioned" or "formed" to distinguish between God's art and that of a mere craftsman, who must fashion his works out of raw materials.[66] For Basil, God "at one and the same time devised what sort of a world it should be and created the appropriate matter with its form."[67] He is not merely "the inventor of the shapes but the creator of the very nature of all that exists."[68]

On the issue of God's creation of matter Basil insists that Plato was wrong. But in most other respects Basil's interpretation of Genesis attempts to incorporate as many features of the *Timaeus* as it can. For example, Basil argues that the term "beginning" refers only to time and the physical world. Something did exist before "the beginning," a world that *is* timeless, without beginning or end, attainable not by sensory experience, but by contemplation alone, one in which the Creator perfected his invisible and spiritual creatures.[69] This world is, of course, closely related to Plato's ideal realm of which the physical world is an image. Moreover, Basil reiterates the Platonic argument that it is because of the likeness between the physical and the transcendent world that the visible world "provides guidance to the mind for the contemplation of the invisible."[70]

Though Genesis does not stipulate that God created some small number of elements to use in fashioning terrestrial and celestial entities, Basil accepts the traditional earth, water, air and fire tetrad, and follows Plato in seeing air and water as means between the extremes of earth and fire.[71] Regarding the Aristotelian presumption that celestial objects are composed of a fifth, ethereal, element, Basil rehearses the arguments pro and con and judiciously argues that in the absence of (1) a scriptural resolution, (2) convincing sensory evidence, or (3) a conclusive rational proof, we must suspend judgement and "stop talking about the substance."[72]

Basil uses the second verse of Genesis, "But the earth was invisible and unfinished" ("without form and void" in the King James translation) to reemphasize the falseness of the preexistent matter doctrine. Earlier commentators, and Augustine, who shows one of his few substantial divergences from Basil on this point, suggest that this passage hints at the existence of unformed matter prior to God's creative act. Basil, on the other hand, borrows from Aristotle in insisting that we must not "search for some nature destitute of qualities, existing without qualities, of itself." "You will end with nothing," he says, "if you attempt to eliminate by reason each of the qualities that exist in it. In fact, if you remove the black, the cold, the weight, the density, the qualities pertaining to taste, or any others which are perceptible, there will be no basic substance."[73] For Basil, verse two of Genesis simply acknowledges that the earth was invisible because (1) there was as yet no observer to view it, (2) it was still covered by water, and (3) there was as yet no light to illuminate it. Similarly, it was unfinished because it had not yet been furnished with its cover of living beings.

The next major issues are tackled by Basil in his third lecture which begins with verse six, "Let there be a firmament in the midst of the waters to divide the waters." In the first place, Basil says that the form of God's address implies that he is speaking to a co-worker; and this allows him to incorporate the Alexandrine "Logos" doctrine in which Christ acts as God's co-worker and as the equivalent of the Platonic Demiurge.[74]

Even more importantly, this passage allows Basil to raise an issue regarding God's omnipotence, which was to be the focus of long-standing and important arguments among Christian scholars. Verse one had already mentioned the creation of the heavens. And some commentators saw verse six as a mere amplification and refinement of the initial statement; for Plato, among others, had insisted that there was but *one* heaven. Yet to Basil, verse six seemed to suggest the creation of another heaven, different from that mentioned in verse one. Basil recalls to mind that earlier Greek cosmologists had suggested a plurality—even an infinity—of worlds; but more critically, Basil wants no *limits* imposed on God's creativity.

> Although they [the proponents of a single heaven] see bubbles, not only one, but many, produced by the same cause, they yet doubt as to whether the creative power is capable of bringing a greater number of heavens into existence. Whenever we look upon the transcending power of God, we consider that the strength and greatness of the heavens differ not at all from that of the curved spray which spurts up in fountains. And so, their explanation of the impossibility is laughable.[75]

His discussion of the firmament also allows Basil to discuss the Platonic-Aristotelian concentric sphere theory of the heavens and to raise an important point regarding scientific method and the importance of sensory evidence. Basil is enough of a Platonist to insist that reason "is much more accurate than the eyes for the discovery of truth."[76] But he was also trained as a physician and displays the physician's real concern for sensory evidence. Faced with the Platonic theory of celestial harmony, according to which each of the planets "gives out such a melodious and harmonious sound that it surpasses the sweetest of singing,"[77] Basil asks for some sensible proof. He reports that the theory's proponents say, "that, having become accustomed to this sound from our birth, we fail to notice the sound through our early familiarity with it and because we have habitually heard it, like men in smithies who have their ears incessantly dinned."

"To refute such subtleties and unsoundness," Basil insists, " . . . is not the practice of a man who either knows how to use time sparingly or has regard for the intelligence of others."[78] While the authority of reason may be "more accurate" than that of the senses, no doctrine that flouts the evidence of the senses can be taken seriously. Indeed, Basil frequently reiterates the notion that sensory experience is extremely important in generating knowledge of the world. He tells his listeners to "employ skill in making an investigation" of the characteristics of different liquids, and disqualifies "he who has not learned by experience" from commenting on botanical phenomena.[79] And he emphasizes that "long experience, by collecting useful information from individual incidents," produces not only utilitarian knowledge, but also an increasing love and admiration of the Creator.[80] Thus he explicitly encourages an important set of scientific attitudes and activities, which he also implicitly adopts throughout his later lectures.

Verse six of Genesis says that the function of the firmament was to divide the waters from the waters; this provides Basil the key to another important substantive aspect of his cosmology. This passage clearly suggests that there is more water present in the initial creation than will later be collected into the seas on the earth. Basil must explain why God should have created so much water in the first place that he needed to move some out of the way. Basil appeals to Isaiah, who prophesies the end of the world in fire, and to the dominant Stoic cosmology which also foresaw the end of a world cycle in a general conflagration. According to Basil,

> The Ruler of the universe ordained from the beginning such a nature for moisture that, although gradually consumed by the power of fire, it would hold out even to the limits prescribed for the existence of the world. He who disposes all things by weight and measure knew how long a time He had appointed to the world for its continuance, and how

much had to be set aside from the first for consumption by the fire. This is the explanation for the superabundance of water in creation.[81]

Some of this water was presumably set above the firmament to be dribbled into the visible part of the universe to replace the waters obviously used up by evaporation under the intense heat of the sun. This superterrestrial water comes down in the form of rain and snow, and is added to the water that falls in the ordinary course of the water cycle of evaporation, condensation, and return to the seas.[82]

Next, in Genesis, God says, "Let the waters below the heavens be gathered into one place." Here Basil faces a fascinating difficulty and opportunity; for as he readily admits, such a statement is superficially "contrary to experience, for all the water does not seem to have run together into one place."[83] At one level we must simply use common sense and admit that while there may be small, separate accumulations, there is one gathering, the connected oceans, which sets the whole element apart from the rest. But we can also interpret God's command in a slightly different way—as establishing the natural law which sets up the water cycle by which each drop of water, whatever its temporary place, ultimately tends to make its way toward the seas.

Here Basil first emphasizes the Platonic notion of God as the source of all natural law: "the voice of God makes nature, and the command given at that time to creation provided the future course of action for the creatures . . . the element of water was ordered to flow, and it never grows weary when urged on unceasingly by this command."[84] This notion is one that Basil frequently reemphasizes. In connection with the command for the earth to bring forth vegetation, for example, Basil writes, "the voice which was then heard and that first command became, as it were, a law of nature and remained in the earth, giving it the power to produce and bear fruit for all succeeding time."[85] In general, Basil asserts the continuance of natural law grounded in God's initial commands.

> As tops, from the first impulse given to them, produce successive whirls when they are spun, so also the order of nature, having received its beginning from that first command continues to all time thereafter, until it shall reach the common consummation of all things.[86]

The constancy of natural law is such an ingrained assumption among most modern Western thinkers that Basil's attitude may seem trivially obvious. But the assumption that there are universal, unchanging natural laws to be discovered in the natural world is both a critical assumption for the very existence of science as we know it, and an assumption that is not shared within all conceptual frameworks. Certain forms of Marxian dialectical materialism, for example, deny this assumption, and many early nature religions allowed for an irregular willfulness associated with nature (see chap. 2). That Christian exegetical writers like Basil adopted the Greek scientific thinkers' notion of natural law and embedded it deep within Christian theology has made it possible for Christian culture to be peculiarly hospitable to scientific thought through much of the last fifteen hundred years.

Though numerous other topics are addressed by Basil, utilizing especially the knowledge presented in Aristotle's works on natural history, there are just two further themes that demand special comment because of their long-term cultural significance.

Verse fourteen of Genesis establishes the existence of lights (stars) in the firmament of heaven and stipulates that they "serve as signs and for the fixing of seasons." This naturally invites some commentary on and criticism of astrological doctrines.

Basil begins his comments on the passage by following Aristotle's *Meteors* in discussing lunar and solar signs that allow weather prediction. Furthermore, he focuses on how useful such signs can be for farmers and sailors. But he attacks those "who go beyond bounds [to] interpret the divine utterance as a defense of astrology and say that our life depends on the movement of heavenly bodies."[87] Again, like Aristotle, Basil does not deny that terrestrial events may be to some degree influenced by celestial ones. Indeed, he calls attention to the solar impact on seasonal phenomena and the lunar influence on the tides.[88] Instead, Basil's most fundamental criticism of astrology derives from the unacceptability of its consequences. Were the astrologers correct, "the great hopes of us Christians will vanish completely since neither justice will be honored nor sin condemned because nothing is done by men through their free will. Where necessity and destiny prevail, merit, which is the special condition for just judgement, has no place."[89] Therefore the astrologers cannot be correct.

In addition, Basil offers some technical criticisms of astrological techniques and interpretive schemes. If, as some astrologers claimed, the influence of the stars at nativity depends on the precise time and the precise positions of the heavenly bodies, no accurate predictions can be made because of observational uncertainties. In theory, astrologers claimed significance in position down to a second of arc, yet naked eye observations are good only to on the order of ten seconds. Moreover, it is extremely difficult to determine precisely the moment of birth. Consequently, as Basil says, "if it is impossible to find the hour accurately, and the change of even the briefest [spatial] interval causes utter failure, both those who devote themselves to this imaginary art and those who are all agape at them as if they were able to know their destinies are ridiculous."[90] Similarly, Basil ridicules the assignment of the characteristics of the animals whose names are associated with signs of the zodiac to those born under the signs.[91] And he criticizes the notion that the influence of a heavenly body should depend on its relationship to other bodies rather than on its own constant and intrinsic God-given properties.[92]

Overall, though Basil shows somewhat less ingenuity than Augustine and others in refuting astrological doctrines, his basic attitudes were widely shared. Celestial objects undoubtedly had some bearing on terrestrial phenomena, but their effects could not be totally determinative—especially in connection with issues related to moral responsibility. Astrologers, moreover, could not fully discover celestial influences in connection with horoscopes because of purely technical limits on recording times and positions. Finally, the specific schemes that associated human characteristics with specific configurations of heavenly bodies or signs of the zodiac could be given no justification in reason or experience. All of these criticisms, of course, left open the possibility that some scheme might be discovered for determining the limited influence of heavenly bodies on phenomena and on character; thus the problem of the existence and degree of astral influence remained very much alive for Christian intellectuals in the medieval world.

Only one fundamental cosmological issue remains to be addressed: the place of man in the cosmos. On this topic Christian cosmologists adopted the Platonic and Aristotelian arguments for the special character of human reason, and sought to expand the difference between man and other natural entities to conform to the Biblical notion that man is not simply a marginally distinguished part of nature, but unique both because mankind alone is made in God's image and because to mankind alone is given mastery or "dominion" over all other living creatures.

In order to emphasize the distinction between man and the other animals Basil produces a unique theory regarding the blood and souls of animals, which draws from a variety of medical and philosophical doctrines. He begins by investing Genesis, verse 24, "Let the earth bring forth a living creature," with new meaning. "Why does the earth bring forth a living creature?" asks Basil. "In order that you may learn the difference between the soul of a beast and that of man."[93] He simultaneously adopts Aristotle's definition of "soul" as that which gives "life" to an organism, and the medical and scriptural notion that "life" is in the blood. From this he concludes that animal souls are blood. But, he argues, because, "the blood when congealed, is wont to change into flesh, and the flesh, when corrupted, decomposes into earth, reasonably, the soul of animals is something earthy. Therefore . . . you will find that the soul of beasts is earth."[94] Human souls, on the other hand, are the direct, immediate, and immaterial creation of God. The whole Platonic theory of metempsychosis must then be abandoned, as must all evolutionary schemes like that of Anaximander who naturalistically discussed man's origins out of a fishlike creature. "Shun the idle talk of the proud philosophers," Basil says, "who are not ashamed to regard their own soul and that of gods as similar, [and] who say that they were at some time women, or bushes, or fish of the sea."[95] In early and medieval Christian cosmology, man must thus be understood as radically distinct from the rest of creation.

Though subsequent early commentators on Genesis diverged from Basil on details and in emphasis—some, like Augustine and Ambrose, for example, were less literal and more allegorical in their approaches—the broad lines of Christian cosmology were set down by Basil. Except for the Christian insistence that God *created* matter as well as *shaped* it, the early Christian universe was basically that portrayed by Plato in the *Timaeus*. It was one created *good* by a rational creator who worked "according to number and measure"[96] and who initially established natural laws to direct the subsequent course of events. The physical world was, in some sense, imperfect; but its study could lead to an understanding and appreciation of its creator. The first physical elements of the world were understood to be earth, water, air, and fire; and these were generally understood to be disposed in concentric rings, beyond which a nesting set of heavens carried the celestial bodies—the moon, sun, planets and fixed stars—which controlled the day, night, and seasonal phenomena, and which might have some influence, but not a determinative one, on human character and events. Beyond the outermost visible heavens lay the waters, and perhaps additional heavens, which God

PLATE 1 The earliest known writing on a tablet from Kish, C. 3500 B.C. (*Republic of Iraq, State Antiquities and Heritage Organization, Bagdad*)

PLATE 2 Clay model of sheep's liver used in teaching divining techniques from the First Babylonia Dynasty, 1830–1530 B.C. (*British Museum, London*)

PLATE 3 The upper portion of the astronomical text from the Selucid Era gives positions of venus from A.D. 180 to 241. The lower portion explains the procedures for calculating positions. *(British Museum, London)*

PLATE 4 Boticelli's *Saint Augustine*. The astronomical instruments and mathematics text in the upper part of the picture link Augustine with scientific knowledge. *(Ognissanti, Florence)*

PLATE 5   Title page from Filipo Calandri's *Pictagoras Arithmetrice introductor* printed at Florence by Lorenzo d Morgiani and Giovanni Theodesco da Manganza, January 1, 1491. The drawing shows Pythagoras teaching mathematics to Florentine artizens. *(Hoover collection, Harvey Mudd College)*

PLATE 6 Page from Calandri's work showing one of the first known illustrated algebra problems. Three master masons agree to build a house. The first working alone would take 10 days, the second 12 days, or the third, 15 days. How many days will they take to complete the house working together? *(Hoover collection, Harvey Mudd College)*

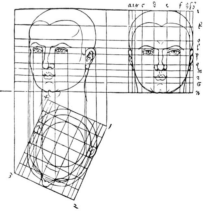

PLATE 7 Page from Albrecht Dürer, *Hierin sind begriffen vier bücher von menchlicker proportion [Four Books on Human Proportions]*, Nuremberg; H. Formschreyder, 1528. This page illustrates Dürer's development of projective geometry for artistic purposes. *(On loan to Sprague Library, Claremont, California, from Michael Wilson)*

PLATE 8 Title page from Georg Andreas Böchler, *Theatrum Machinarum Novum*, Nuremberg, 1662. This plate from a typical Renaissance machine book clearly depicts the fusion of theory with practice or study with labor which was emphasized by early modern architects and engineers. *(Hoover Collection, Harvey Mudd College)*

A—FURNACE. B—ITS ROUND HOLE. C—AIR-HOLES. D—MOUTH OF THE FURNACE. E—DRAUGHT OPENING UNDER IT. F—EARTHENWARE CRUCIBLE. G—AMPULLA. H—OPERCULUM. I—ITS SPOUT. K—OTHER AMPULLA. L—BASKET IN WHICH THIS IS USUALLY PLACED LEST IT SHOULD BE BROKEN.

PLATE 9   Chemical apparatus used in metallurgy. From Georg Agricola *De Re Metallica* (1556) translated by Herbert Clark Hoover, London; *The Mining Magazine*, 1912. *(Hoover Collection, Harvey Mudd College)*

PLATE 10 Illustration of the macrocosm/microcosm analogy, with lines drawn from signs of the zodiac to regions of the human body that they influence. From Robert Fludd, *Ultrisque Cosmic Maioris scilicet et Minoris Metaphysica, Physical atque Technia Historia . . .* Oppenhem, 1617. *(The Francis Bacon Library, Claremont, California)*

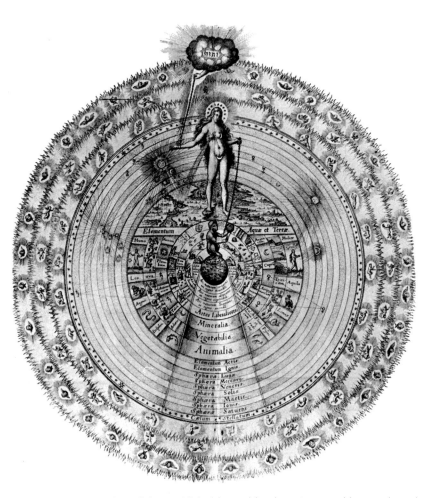

PLATE 11 Depiction of the multiple hierarchies in nature and human knowledge from Robert Fludd, *Ultrisque Cosmi. . . . (The Francis Bacon Library, Claremont, California)*

in his infinite power might have chosen to make. God's wisdom was particularly evident in the design of the creatures that inhabit the earth, and useful moral lessons could be recognized in the behavior patterns of various creatures. Within this creation God set man apart—a being capable of reason and moral free choice—and therefore the only creature capable of being good or evil. Moreover, God assigned to this creature dominion over the rest of his creation.

There is no doubt that western Europe between A.D. 100 and 1000 had no recognizable scientific specialty, and that it saw a dramatic decline in levels of scientific knowledge. But if the level of scientific knowledge incorporated into early Christian theology was relatively low, a crucial issue remains irrefutable. In early Christian culture the all-important religious specialty incorporated many of the domains of scientific thought as well as many of the methods and attitudes of the old Greek scientific specialty. What the long-term consequences of this integration were, we will recurrently consider.

SCIENCE AFTER THE FALL OF ROME

During the early centuries of the Christian Era, there was never any question, even in the most learned and scientifically oriented Christian writings, that secular concerns were subordinate to religious ones. This is illustrated by the very form of the Hexamera, in which the commentary is on Genesis rather than on the world. But by the middle of the twelfth century a subtle, yet critically important shift occurs. This shift is symbolized by the departure from the traditional form in the Hexameron by Thierry of Chartres. As Brian Stock has emphasized, in two of the three sections of Thierry's work, "Instead of following the traditional method of fitting natural philosophy into the historical framework of the Bible, [he] fits the opening chapters of Genesis into the framework of natural philosophy."[97] The possibilities foreseen by Tertullian and Lactantius had come to pass, and the way had been opened for the scientific tail to wag the theological dog. How this set of circumstances came about is the focus of this section; with what consequences to medieval culture is the focus of the next chapter.

The intellectual and social peak of early Christianity occurred just before the effective destruction of the Christianized Roman Empire. After nearly three centuries of gradual growth, predominantly among the lower classes and among women of all classes,

Christianity was officially embraced by the Roman Emperor, Constantine, in A.D. 312, bringing on the conversion of wealthy and powerful citizens and the support of a line of Emperors. From a frequently persecuted minority cult it became the dominant religion and, in A.D. 352, the only legal religion in Rome. At the same time, upper-class conversions brought in able, classically trained intellectuals like St. Augustine, St. Ambrose, and St. Jerome, who produced critical linkages between Christian doctrines and Greco-Roman learning.

But just as Roman Christianity began to take on "respectability" and some sophistication, the society to which it was linked in western Europe entered into a period of rapid and dramatic decline. Perhaps most important for our story, the western empire became depopulated and almost completely deurbanized as a result of economic pressures and political ineptitude. By the beginning of the fifth century the once vigorous urban centers of Italy and northern Europe, drained of their wealth and population to support the Roman armies and bureaucracy, had all but disappeared; and the few families to retain any wealth had retired to their country estates to evade taxes and to build up local protective armies. A new rural elite with nothing but local and agricultural interests was emerging.

At the same time, a whole group of vigorous, illiterate, war-oriented tribes from the North invaded the western empire and carved it up. In the Far West, the Franks, Angles, and Saxons took over the regions now occupied by England, France, and the Netherlands. Initially unchristianized agrarian peoples, these groups moved in slowly, but once settled, they hung on tenaciously. The regions we now know as Spain, Italy, Southern Germany, Switzerland, Austria, and North Africa were overrun by the more nomadic Ostrogoths, Visigoths, and Vandals, who were soon nominally Christianized, or who at least tolerated Christianity; but civilization as it had been known for at least a millennium in the western Mediterranean was effectively destroyed.

Rome was first sacked by the Visigoth, Alaric, in 410. St. Jerome voiced the sentiment of many when he lamented, "My tongue sticks to the roof of my mouth, and sobs choke my speech."[98] Just twenty years later, in the year of St. Augustine's death, the Vandals overran his North African city of Hippo.

For a brief period, from A.D. 493 to 526 under the relatively Romanized Ostrogothic king, Theodoric, there was a minor revival of Roman-Christian culture. Boethius, a Christian high official in

Theodoric's regime, translated and reintroduced the study of the logical and methodological writings of Aristotle into the educational scheme, and wrote a critical work on Greek mathematics, the *Arithmetic*. In addition, he produced one of the most widely read medieval works, *On the Consolations of Philosophy*, which urged Christian consideration of pagan philosophers—especially Plato and Aristotle. Cassiodorus, another Christian scholar and Roman aristocrat who was Theodoric's secretary and a great admirer of Boethius, had perhaps a greater long-term impact on Western learning. Worried about the state of religious learning in the Latin west, Cassiodorus not only wrote a compendium of knowledge, *On Training in Sacred and Profane Literature*, but he also founded a monastery on his estates, gathering together a group of monks whose primary aim was to accumulate manuscripts and to translate non-Latin works from the Greek and Hebrew traditions. In particular, he focused consideration on the so-called seven liberal arts as a background to Christian theology. Thus Cassiodorus established the institutional context within which virtually all Christian learning was focused until the emergence of universities in the twelfth century.

When St. Benedict founded his vastly more important monastery at Monte Cassino a few years later, he incorporated Cassiodorus' educational concerns into what was for nearly seven centuries the central institution of religious and intellectual life in western Europe. Benedictine monasticism, like all monasticism, embodied a fundamental impulse to renounce and withdraw from the secular world to seek a more perfect Christian life. But Western monasticism had an incredible impact on the society it formally renounced because Benedictine monks (and the independent Irish monks) produced virtually all literate Europeans during the early Middle Ages. They transcribed and preserved whatever classical learning there was. They provided the missionaries who gradually Christianized all of northern and western Europe. They served as clerks and advisors to kings, gradually increasing in power as medieval society became more complex and was bureaucratized. They were drafted into high church offices. And as the custodians of gifts of land over a long period of time, they managed the large estates that became the centers of technological innovation in Europe, as well as important centers of political and legal administration.[99]

Within the education provided by the Benedictines, the level of scientific sophistication was substantially lower than that evidenced by Boethius or by his predecessors like Augustine or Basil;

but some interest was maintained in scientific topics. And there is evidence from the seventh and eighth centuries that within the Irish monastic tradition a slightly higher degree of interest in mathematical and astronomical topics had been maintained. Thus, when the opportunity to improve the level of knowledge of the subjects of the quadrivium presented itself, beginning in the eleventh century, Christian scholars were eager to take advantage of it.

Natural Science and Theology in the Medieval University, About A.D. 1050−1350

6

Though the Frankish king Charlemagne provided a minor stimulus to education at the end of the eighth century, there were no radical changes until the period of the High Middle Ages—about 1050–1300. Then two factors led to a marked increase in secular—especially scientific—aspects of learning. First, the eleventh century reconquest of Spain and Sicily from Islam brought Christian scholars into intimate contact with a culture that had not merely embraced much of the highest Greek scientific and philosophical learning, but had extended it in many domains—especially in mathematics, astronomy and/or astrology, and medicine. Second, the late eleventh and twelfth centuries saw a spectacular repopulation, urbanization, and commercialization of Europe. This trend produced the characteristic urban need for bureaucrats and for trained professionals in law and medicine as well as for a better educated clergy, and this need in turn gave rise to the institution of the medieval university in which the role of scientist was recreated and given substantial influence in Western culture.[1]

Edward Grant has nicely summarized the consequences of these two trends, which we look at in more detail in this section.

> By the mid-thirteenth century, the arts courses required for the degree of Master of Arts, a degree that was prerequisite for study in the higher faculties of law, medicine, and theology, were heavily oriented toward logic and natural science. . . . The required program of study . . . at the universities of Paris and Oxford was not, as those unfamiliar with the Middle Ages may suppose, top heavy with courses in theology and metaphysics. Rather, it consisted for the most part of courses in logic (which had absorbed much that was grammatical), physics, which embraced physical changes of all kinds, cosmology, and elements of

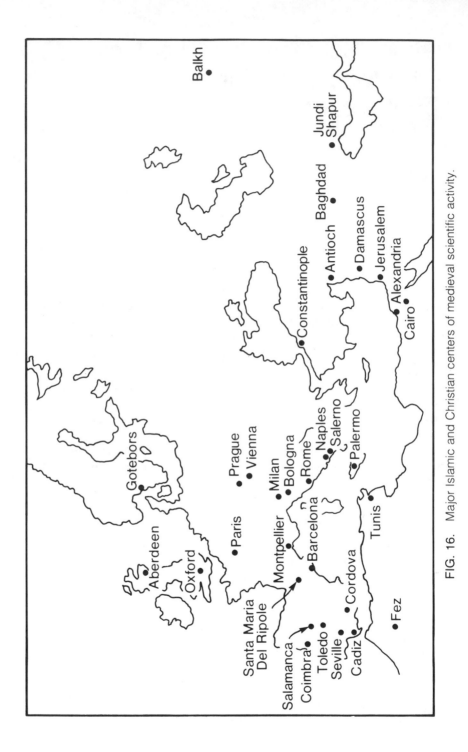

FIG. 16. Major Islamic and Christian centers of medieval scientific activity.

astronomy and mathematics. . . . . *Never before and not since have logic and science formed the basis of higher education for all arts students.*[2]

In order to understand the impact of Islamic science in eleventh- and twelfth-century Christian universities, we must very briefly describe the development of science in early Islamic culture.[3]

THE ISLAMIC BACKGROUND TO MEDIEVAL EUROPEAN SCIENCE

During the seventh century the religion of Islam emerged in Saudi Arabia, spreading during the next three centuries throughout the Near East, North Africa, and into Sicily and Spain. For reasons much too complicated to analyze here, Islamic scholars from about A.D. 800 to 1200 became fascinated by Greek philosophy and science, integrating it with new mathematical learning that came from the eastern part of the empire. The vast bulk of the works of Plato, Aristotle, physicians like Hippocrates and Galen, astronomers like Hipparcus, Aristarchus, and Ptolemy, and mathematical physicists like Archimedes, were translated, commented on, and used as the basis for extended works in mathematics, astronomy and/or astrology, medicine, optics, and alchemy; the studies began at the famous *House of Wisdom* founded at the end of the eighth century at Baghdad under the caliphate of Harun-al-Rashid. By the mid-tenth century the centers of political and intellectual life in Islam had moved westward and northward into Spain and Sicily, where at such centers as Cordova, Toledo, and Palermo, huge libraries or centers of learning much like the first great center at Alexandria came into existence.

As in Hellenic Greece, in Islam during its period of high scientific culture, scientific systems were closely associated with individual wise men and great teachers. To give some sense of the shaping of European attitudes to Islamic science and its classical sources we focus on three specific figures, Abu Ma'shar or Albumasar (ca. A.D. 786–866), who represents the astronomical and/or astrological tradition; Ibn Sina, or Avicenna (A.D. 980–1037), the greatest of Islamic medical writers and encyclopedists; and Ibn Rushid, or Averroes (A.D. 1126–1198), the greatest of Islamic commentators on Aristotle, known as "The Commentator" to the Latin West.

The importance of astrology in stimulating Christian interest in Islamic science can hardly be overestimated. One recent survey of early (eleventh and twelfth century) translating activity shows that nearly 30 percent of all translations from Arabic to Latin were specifically astrological, and that nearly 50 percent involved either astrology, astronomy, or mathematics (which was in large measure utilized for astrological purposes). Just why astrological topics commanded so much interest we do not know. In part, it was probably because they were clearly relevant to an ongoing twelfth-century Christian debate about the extent of human free will and responsibility, and in part it was probably because astrology played a powerful role within the Islamic intellectual tradition itself; Islam was generally inclined to place little importance on human free will relative to the will of Allah, which determines all. Thus, omenology of all kinds, and astrological predictions in particular, were considered to be extremely important.[4]

Of the early astrological texts, Abu Ma'shar's *Great Introduction to the Science of Astrology* and *Little Introduction to the Science of Astrology* were among the few texts of which multiple translations were made, and they were very influential in twelfth-century European intellectual life.[5]

Abu Ma'shar's works beautifully illustrate a number of characteristics of the Islamic sources that became available to Christians in the eleventh and twelfth centuries. Ma'shar came from the ancient city of Balkh (now in the southern Soviet Union, just north of the border with Afghanistan) where groups of Jews, Nestorian Christians, Manichean Gnostics, Zoroastrians, Buddhists, and Hindus, as well as Muslims lived together. Thus Ma'shar had available to him (and used) a wide variety of sources in developing his astrological theories (sources that came down in at least five languages, Greek, Sanskrit, Pahlevi, Syriac, and Arabic). To aid him in synthesizing elements from these diverse and often inconsistent sources he applied, and was one of the prime early advocates of, a theory that became almost universally accepted among Islamic scholars. According to this theory, all systems of thought are ultimately derived from Allah's revelation; thus all contain elements of truth amid the accrued errors introduced by human interpreters.

In broad outline, Ma'shar's philosophical justification of astrology derives from Aristotle's *De Caelo*, though it probably came secondhand. Ma'shar posits three levels of being, like three concentric rings. The first and outermost is the sphere of the divine; the second, ethereal, sphere contains the paths of the stars, planets, sun and moon; and the third, innermost, sphere is the locus of the four transmutable and moveable traditional terrestrial elements. The whole system is unified by the fact that the outer (superior) spheres influence events in the interior (inferior) ones. Terrestrial bodies have a potential to be moved (i.e., changed) by the influence of particular celestial bodies, and celestial bodies have the potential to influence specific terrestrial ones. The major implication of this theory along with an important medical parallel is nicely summarized in the *Great Introduction*:

> The object of astrology . . . is [first] to measure the motions of the elements from the motions of the stars and the changes of time; then to measure the motions of the cosmos itself; then, of its parts, [and to do these] both generally and specifically. Just as the physician is first instructed by the sensations of his experience, then turns to the natural properties of the species . . . , so the astrologer turns from a certain sensible instruction of experience to the natural [relations] of heavenly bodies. . . . And thus he recognizes the definite power and natural force of the stars and planets by reason of their effects . . .
>
> Medicine, starting from their composition, status, and accidents, deals with the nature of the elements and the changes of bodies; but astrology, as a whole, reaches its consummation in the motion and natures of celestial bodies and their effects throughout the lower world. The physician pays attention to changes in the elements; the astrologer follows the motions of the stars in order to understand the causes of elemental change.[6]

Insofar as this passage deals with astrology, it is wholly consonant with Aristotelian doctrine; but in its linkage between astrology and medicine it extends Greek doctrines in a way that was to be increasingly important in Islam and Europe. According to Abu Ma'shar, the physician can be effective in treating patients because he learns how mixtures of different materials cause changes in bodies—in the human body in particular. Since astrological knowledge [supposedly] allows one to know how terrestrial bodies interact, astrology should be a useful adjunct—perhaps even a prerequisite—to effective medical practice because it can provide knowledge of the circumstances under which certain changes can be produced by specified mixtures. It is this set of assumptions that give rise to an intimate linkage between astrology and medicine throughout Islam and Europe well into the Renaissance. Note also that Ma'shar's justification of astrology is totally scientific in its attitudes. However much he may have failed to limit his schemes to purely phenomenal elements and to testable statements, his goal was *verifiable systematic knowledge of the universal causes of phenomena.*

To his philosophical and scientific justification of astrology derived from Aristotle, Ma'shar adds a peculiar religious doctrine adapted from Zoroastrian and Gnostic elements. Human souls come into the world by descending from the divine region through the celestial regions, into the terrestrial sphere. They must strive to return to the divine, but they cannot simply transcend the celestial region without help. Instead, they must return with the help of the intermediate deities, the celestial bodies. Religious rituals must be addressed to these intermediate deities, whose attributes and powers are known through astrology and astronomy.

Note here that the Platonic vision of the celestial bodies as intermediate deities and governors of the world is central and that the older Gnostic vision of these deities as fundamentally evil has been transformed into one in which the celestial deities can be helpful as well as harmful to man's spiritual and material well being. Moreover, though the general doctrines may initially seem incompatible in the extreme with both Islamic and Judeo-Christian monotheism, Ma'shar and many of his Islamic and Christian interpreters insisted that they could be sustained within orthodoxy by constantly focusing on the subordination of the celestial deities to the will of the single transcendent God. Not only were these doctrines widely supported within Islamic theology, but for a time under the Renaissance Pope Alexander VI they were specifically declared to be in conformity with the Catholic faith.

When we turn from general philosophical and theological rationales for astrology to the specific technical details of forecasting schemes, we find that Abu Ma'shar continues his eclecticism. His general theory for determining planetary and stellar positions is based on the Greek Ptolemaic model of the universe, but mean motions of planets were calculated from Indian parameters; and other parameters, including the prime meridian, were taken from Persian sources. Furthermore, Ma'shar offered a general astrological interpretation of history that was Zoroastrian, while he used Hellenistic Greek treatises for his theory of individual horoscopic astrology.[7]

Though Abu Ma'shar's technical schemes probably stimulated European interest in mathematics and Ptolemaic astronomical theory, it is fairly clear that his philosophical and theological views were of greater general importance. In particular, his attitudes were incorporated into the works of the widely read twelfth-century literary figure, Bernard Silvester.[8] Moreover, when Aristotle's works became available in the thirteenth century, the close parallels between the arguments of *De Caelo* and those of Ma'shar intensified secular interest in (and theological suspicion of) Aristotle.

If the precise reasons for Latin interest in Islamic astrology are problematic, it is easy to understand the reason for Latin interest in Islamic medicine and in the greatest of Islamic medical works, the *Canon of Medicine* of Avicenna. Islamic medicine, which had from its very beginnings incorporated virtually all of the Hippocratic Corpus and the works of Galen, was world renowned for its excellence. When Sicily came under Norman rule in the eleventh century, and Latin speakers regularly came into contact with Islamic medical texts at Palermo, it was natural that those texts be translated for use in Latin medical education. Of all Islamic medical men the most famous was Avicenna, author of over two hundred and fifty works, including *The Canon of Medicine*, which became one of the most frequently printed scientific works of the Renaissance and remains the foundation of Near Eastern medical training even today. Avicenna's work is too technical and specific to be considered extensively here; its excellence often lies in the careful description of diseases and drugs through case histories. For our immediate purpose his greatest cultural impact can best be understood by considering part of a description of Avicenna's importance by the twelfth-century Islamic historian Niẓāmī Arūdī.

> For four thousand years the wise men of antiquity travailed in spirit and melted their very souls in order to reduce the science of Philosophy to some fixed order, yet could not effect this, until after the lapse of this period, that incomparable philosopher and most powerful thinker Aristotle weighed this coin in the balance of Logic, assayed it with the touchstone of definitions, and measured it by the scale of analogy, so that all doubt and ambiguity departed from it, and it became established on a sure and critical basis. And during the fifteen centuries which have elapsed since his time, no philosopher hath won to the inmost essence of his doctrine, nor travelled the high road of his method, save that most excellent of the moderns, the Philosopher of the East, the proof of God unto his creatures, Abū 'Ali al-Husayn ibn 'Abdu' elāh ibn Sīnā (Avicenna). He who finds fault with these two great men will have cut himself off from the company of the wise, placed himself in the category of madmen, and exhibited himself in the ranks of the feeble minded.[9]

Nizāmī Arūdī's comments show the extent to which Avicenna had associated his general framework of argumentation with Aristotelian philosophy, and the extent to which this association heightened the reputation of Aristotle. Before the beginning of the eleventh century, Aristotle had been widely translated and used by Islamic thinkers in their syncretistic works. In general his reputation was comparable to that of Plato and Pythagoras, but certainly not set above them. Indeed, earlier Islamic scholars had shown vastly greater interest in the mathematical orientation of Plato and the number mysticism of the Neoplatonists than in the empiricist cast of Aristotelian thought. But medicine was the natural home of empiricism, and when a truly outstanding scholar and physician adopted and wholeheartedly expounded Aristotelian ideas, his stock rose dramatically among Islamic thinkers. Moreover, Avicenna's support for Aristotle could not have been lost on his Latin readers, and it was shortly after *The Canon of Medicine* was translated that many of Aristotle's works began to be translated from the Arabic into Latin.[10]

Though interest in Aristotelian science was undoubtedly helped along by Avicenna's works, the very special and controversial place of Aristotelian science in European intellectual life during the High Middle Ages is most closely associated with the purest Muslim Aristotelian, Averroes, who spent most of his life at Cordova becoming an authority on religion, law, and medicine as well as Aristotelian philosophy. Much of Averroes' fame or notoriety probably derived from a misunderstanding of his religious attitudes among Latin scholars, but that can in no way detract from his significance.

In his thirty-eight commentaries on Aristotle's works, Averroes provided a detailed explication of the entire range of Aristotelian thought. His ideas on sublunary topics demand no comment because they faithfully follow Aristotle; but his attitudes toward the relationship between philosophy and religion demand consideration because they stimulated a renewed outbreak of the conflict between the Christian defenders of faith, and of reason, which had been temporarily set aside by Augustine.

Like Aristotle, Averroes was convinced that human reason, working on sensory experience, was capable of generating all possible knowledge, including that of the existence of the Divine. In one sense this conviction was little different from that expounded by Origen seven centuries earlier; but there were some important perceived divergences. Though Averroes was personally convinced that reason and revelation are simply different sources of the same divine truth, his method of discussing this belief was couched in such a way as to appear to Christian readers like an attack upon theologians by rationalistic philosophers, and upon faith by reason.

The basic focus of Averroes' comments on the relation between science and religion or, more accurately, between philosophy and theology, was a theological attack upon Avicenna's Aristotelianism, *The Destruction of the Philosophers*, by the Islamic theologian al-Gazali. To this work, Averroes responded with a *Destruction of the Destruction . . .* and *The Agreement of Religion and Philosophy. The Agreement* is an easier and clearer text to summarize, and while it was probably not known in Latin translation until the fourteenth century, there is no question that it had a substantial thirteenth-century underground reputation among Latin scholars.

As a shrewd analyst of Islamic religious law, Averroes began his discussion by pointing out that the Koran specifically stipulates that men should seek knowledge of their maker through study of nature and its interpretation by reason.[11] Since for Averroes, Aristotle had conclusively established the methodological and metaphysical grounds for any such study, it was unthinkable that there could be any real conflict between Aristotelian philosophy and Islamic revelation. How, then, did the apparent conflicts arise? Here Averroes developed an argument that had been hinted at by Clement and Origen and put it in an Aristotelian framework. Clearly, not all humans are capable of following the complexities of detailed philosophical arguments; so some alternative forms of religious instruction must be available, suitable to their needs and not subject to the rigorous demands of philosophical truth.

Averroes follows Aristotle in distinguishing three levels of argument suitable to three kinds of people—rhetorical, dialectical, and necessary arguments; suitable for the common man, the theologian, and the philosopher respectively.

For the common man, religious injunctions must be embodied in an imaginative literature that will appeal to his passions, and to which he can respond with little or no thought. Such men believe on the basis of faith, and the revelation granted to them may not contain truths in the strictest sense. Here we see a clear variant on Clement's and Origen's demands that texts like the Bible be interpreted allegorically to save them from absurdity.

So far, so good. But now Averroes takes a very dangerous second step. Some people demand more than simple appeals to the imagination. They want to be assured that what they are asked to believe is plausible, or consonant with established scientific knowledge and common sense. It is the function of theology to create plausible arguments for beliefs, which such men are asked to hold by *faith* (i.e., without rational grounds). Such arguments will generally *not* be real demonstrations, but merely psychologically compelling reflections. They are important; for without them a large and important class would be deprived of faith and left without a substitute. But such arguments are *not* philosophically acceptable.

Averroes' true philosophers, like Clement's Christian Gnostics, and certainly like Aristotle's contemplative or theoretical philosophers, are capable of something more, and demand more. These men constitute the intellectual elite and they demand to know both what is, and why it is so and not otherwise. Generally there will be no real conflict between theologically justified belief and rationally grounded truth, for each is the property of a different class of minds, and both are derived from the One Divine. But occasionally what theologians find plausible, philosophers know to be impossible. Then the philosophers cannot be asked to give up their knowledge. Nor, however, should they be so insensitive as to insist that theologians give up their faith. Consider a single important example. Aristotelian philosophy demonstrates the eternity of the world—that is, that it could not have had a beginning. Theologians, on the other hand, point to revealed texts and plausible arguments to convince themselves that the world was created in time. Averroes acknowledges that neither group could with integrity abandon its position; thus he is forced to advocate the maintenance of a truce through establishing an esoteric, secret, and publicly forbidden philosophical tradition.

That such an argument should be misinterpreted as an overt attack on revealed religion, and that it should have some appeal to philosophers but be found offensive to theologians within both Islam and Christianity can hardly be surprising.

Had Averroes' notions about philosophy and religion appeared in Latin Christendom in the tenth century, or without their intimate links with his commentaries on Aristotle's works in natural philosophy, they would probably have had almost no sympathetic hearers and therefore no significant impact. But by the early thirteenth century, when they became known, there was a dedicated coterie of students and teachers of Aristotelian natural philosophy within the medieval European universities.

THEOLOGIANS VERSUS THE ARTS
FACULTY IN THE MEDIEVAL UNIVERSITY

Until the end of the eleventh century almost all education available in Christendom was provided in monasteries and cathedral schools and was aimed at religious instruction—though secular ends were incidentally served. In addition there were some lay schools that taught liberal arts at an elementary level and Roman law at a slightly higher level in the few larger northern Italian cities. With the increasing late medieval urbanization and complexification of society, these schools proliferated and specialized. At Salerno and Montpellier, for example, the need for physicians and the availability of new Islamic medical knowledge led to the development of substantial, but largely informal, medical schools in the twelfth century. Similarly, increasingly complex interactions between the Papacy and secular leaders, as well as increasingly complex commercial activities led to the growth of widely known schools of Roman and canon law, first at Bologna, Italy, in the twelfth century, and then in several urban centers. Moreover, increasing demands for learned clerics led to the growth of some older cathedral schools, like that at Notre Dame in Paris, into larger centers of theological study. These schools grew up in different ways, but by the early part of the thirteenth century an important trend was under way, supported in some places by student pressures and at others by papal support. At the larger institutions, like those at Paris, Oxford, and Montpellier, the traditional higher faculties of medicine, law, and theology were joined by a more or less autonomous faculty of arts whose function was to teach the trivium and quadrivium to those who wanted to enter the higher schools.[12] Such collections of masters and students became known as universities. At most institutions the arts faculty was substantially larger than the others because the arts courses typically became a formal prerequisite for all higher study.

Initially, it is almost certain that arts courses were taught by men whose central intellectual commitments were theological, legal, or medical; but a set of circumstances led rapidly to the emergence of professional arts masters. In almost all universities students paid the arts masters directly. In the Italian universities, masters served at the students' pleasure. At Paris and other places, Papal encyclicals directed that all qualified (degree holding) candidates be freely licensed to teach.[13] Thus a market mechanism was established that placed a premium on excellence in arts teaching, and made it possible for an outstanding teacher to earn a good living by specializing in some aspect of the quadrivium.

By quite early in the thirteenth century, then, the medieval university gave rise to a relatively large class of men whose specializations were in secular learning—especially in the topics of the old quadrivium (mathematics and natural science)—and who were therefore intensely committed to incorporating the newly recovered Islamic and Aristotelian scientific learning into their teaching.

In the long run this rise of secular learning provided the foundations of the Western scientific tradition as we know it. But in the short run it produced (1) an intellectual and institutional power struggle between theology and science, which was a central *focus* of thirteenth-century intellectual life, and (2) the rapid spread of Aristotelian scientific and philosophical perspectives, which was the central *fact* of intellectual life in the High Middle Ages.

The two greatest medieval universities, those at Paris and Oxford, had begun primarily as theological schools dominated in curricular matters by the faculties of theology. And at Paris in particular the rise of a powerful arts faculty generated substantial opposition. Thus, at Paris there was a constant tension between the theologians and arts masters, which led to at least three official prohibitions of Aristotle and his Arab commentators in the first third of the thirteenth century, in 1210, 1228 and 1231. At Oxford the leading theologian and intellectual, Robert Grosseteste, embraced the new learning, leaving Oxford free to become the leading center of scientific study in the thirteenth century. Even at Paris, however, the need for repeated prohibitions of Aristotle and the commentators indicates the existence of a substantial unofficial interest in them.

By 1231, the Paris ban on Aristotle's naturalistic works directed that a commission of theologians edit them, "so that, what are suspect being removed, the rest may be studied without delay and without offense."[14] And by 1255, virtually all of Aristotle's works on natural science, including the *Physics, On the Soul, On the Heavens, On Generation and Corruption* and *On Plants* had been added to his logical works as the official subjects of lecture courses and examinations in the arts faculty. As Gordon Leff remarked,

> The important thing is that Aristotle had now effectively become the arts course. . . . Aristotle and Procius and the Arabian systems based on them had no connection with Christian conceptions; they did not

represent doctrinally neutral facts like the principles of grammar or indeed logic. They were a self-contained and potentially antithetic world view which at best was independent of Christianity and at worst challenged it.[15]

In 1277 the theologians at Paris struck back for one last time. We return to this conflict in the next section, but first we turn from Paris, where science and Aristotle became virtually identified with one another, to look briefly at some other science-related trends. Even at Paris, as Leff points out in the statement above, the *Book of Causes* by the Neoplatonist Procius was widely used (though the Parisians thought the work was Aristotle's); elsewhere Aristotle, while important, was not totally dominant, and this made for a slightly different kind of science and a dramatically diminished conflict between theologians and the emerging natural philosophers. In general, at Oxford and elsewhere outside of Paris, Arabic mathematics, astronomy and/or astrology, and alchemy, as well as the newly retranslated *Timaeus* of Plato and *Almagest* of Ptolemy, played a more significant role. Since such works tended to conform more closely to the Platonic and Neoplatonic attitudes dominant in Christendom and embodied in the works of Augustine and Basil, for example, they were in some sense less wrenching. At the same time they were much more mathematical; and it was not until the late fourteenth century that Parisian scientific work attained the mathematical sophistication of that at Oxford.

One further characteristic—an ironic one at that—of medieval Aristotelianism and science in general demands some note. The early Christian Hexameral context of scientific thought provided a textual orientation and did not encourage direct experience of phenomena. The medieval university context encouraged this textual and authoritarian bias among those who came to study natural philosophy because the subjects of examination were specific texts and doctrines. Oddly enough, this textual emphasis was nowhere more dominant than in the arts faculty at Paris, where the doctrines of the greatest empiricist and observer of antiquity reigned supreme. At Oxford, where much stronger—supposedly mystical and anti-empirical—Platonic and Neoplatonic traditions of mathematics existed, we begin to see the emergence of a renewed experimental—or at least, observational—science in the writings of Grosseteste and his student Roger Bacon, among others.[16] This set of circumstances helps to explain why the late Renaissance flowering of empiricist attitudes is often so self-consciously anti-Aristotelian, and is associated with a revitalized Platonic and Neoplatonic tradition.

THE CONDEMNATIONS OF 1277 AT PARIS: BACKGROUND AND EVENTS

One particular episode, the condemnation of a set of 219 theses—most of them derived from Aristotle and the commentators Avicenna and Averroes—by Etienne Tempier, bishop of Paris, in 1277, offers an opportunity to explore the central issues at stake in the interactions between scientific and religious ideas in the High Middle Ages. Whether the condemnations themselves had any great impact is a matter of vast disagreement among scholars. But there is no doubt that the condemnations were symptomatic of a series of intellectual crosscurrents, and that they serve as a convenient symbol for a set of major issues.

At least until the second third of the thirteenth century, Platonic and Neoplatonic metaphysics and cosmology provided the basic framework for Christian thought. Aristotelian logic and intellectual categories were certainly used, but they were fit into a cosmological and theological system that (a) accepted the notion that universal and eternal ideas existed independent of objects of sense experience; (b) accepted the notion that the physical universe was created in time by God; and (c) insisted that all truths, however they are sought, ultimately derive from a divine illumination rather than from some process of induction from experience.

The rise of Aristotelianism dissociated from its Platonic and Neoplatonic trappings, and reinvested with the trappings of Averroistic thought, led to a wide spectrum of attitudes among the arts faculty and theologians alike. We must sample these attitudes before getting to the condemnations of 1277.

Attitudes toward Aristotle must be understood in the context of the general rising importance of philosophical knowledge relative to theological, legal, or medical studies. Sermons preached by the theologians at Paris complained that, "It is deplorable that the faculty of theology, which is called the republic of sole truth and understanding, should have to speak the language of the philosophers."[17] Similarly, those who were particularly faith-oriented loathed the " 'pernicious men' who sought to reduce the ineffable mysteries of the Trinity, transubstantiation, and other theological truths to 'our understanding . . . and presume to formulate them according to certain natural and philosophical and logical reasons, seeking to include within the rules of nature what is above all nature.' "[18]

Seldom can we find a more direct and clear-cut statement of opposition to the "scientization" of religion than in these two statements of opposition to the encroachment of scientific language and naturalism on the domain of the presumedly supernatural. To combat this trend the Dominican order regulated the study of its brothers in 1228: "They shall not study in the books of the Gentiles and the philosophers . . . They shall not learn secular sciences nor even the arts which are called liberal . . . but they shall read only theological works whether they be youths or others."[19] Curiously enough, while this antiphilosophical perspective had precedents in the early Latin fathers, it was a new perspective in the early thirteenth century. Since Augustine there had been little question that the liberal arts in general, and logic in particular, were necessary prerequisites for theological discourse.

Ironically, it was the Dominican, Albert the Great (ca. 1193–1280), who turned the initial Parisian antiphilosophical attitude toward a more sophisticated analysis of the relationship between theology and philosophy. Albert was simultaneously a theologian and a student of philosophy, a teacher in the theology faculty at Paris from 1243 to 1248, and a teacher of philosophy in the latter part of his life. His first significant move was to emphasize that natural philosophy is a self-sufficient discipline, not dependent on theological applications for its existence or significance. In doing this he emphasized what was becoming an institutionalized reality in the universities, that is, that theology and natural science are indeed different kinds of activity capable of being carried on independently. The one is oriented toward sensory evidence and logical demonstration and seeks to understand only the natural world. The other is oriented toward articles of faith and the world of spirit. Reason may be incapable of demonstrating certain revealed truths; but by the same token, theological dogmas are of little value regarding those subjects open to sensory experience. In terms of our discussion in chapter 1, Albert made explicit the implicitly different goals, attitudes, and methods of the theological and natural philosophical specialties as they existed in the thirteenth century. Albert remained theological in his ultimate commitment, insisting that whereas Aristotle could be wrong, God could not; but his careful distinctions provided the framework within which subsequent arguments developed.

If we admit that science and theology are different in aim, method, and sources, then no problem can arise between them unless they generate contradictory statements about some common concern. In principle one could claim that no conflict could ever arise because there are no common concerns; but to my knowledge no thirteenth-century medieval thinker made precisely that claim. Rather they took up roughly four positions, all of which accepted the validity of philosophy when it did not involve any apparent conflict with theological principles and vice versa. Most theologians, like Albert, following Basil and Augustine, simply asserted the primacy of dogma, faith, or revelation when any conflict existed. But three other positions were probably represented.

The two most radical positions were associated with the so-called Latin Averroists in the arts faculty, who were represented by Siger of Brabant (ca. 1235–1284) and John of Jandun (1286–1328). According to such men, either there are two kinds of truth, so that while something might be theologically true it can be philosophically false and vice versa,[20] or, when philosophical and theological statements conflict, the philosophical statement must be accepted as true. In fact, it is often very difficult to figure out which of these two positions any Latin Averroist held. John of Jandun, for example, was particularly fond of demonstrating the necessary truth of some proposition contrary to theological dogma, then announcing his commitment to the theological belief in the following way: "I do believe that that is true; but I cannot prove it. Good luck to those who can!" Or, "I say that God can do that, but how, I don't know; God knows."[21] While formally orthodox, such ironic statements seemed to have the intent of denying the truth of the assertions they ostensibly upheld.

The final position—that which in the long run became the official position of the Catholic Church, operative even today—was developed by Albert's pupil, St. Thomas Aquinas, who studied and taught theology at Paris off and on from 1245 to 1274. Though Thomas admitted that the methods of theology and natural science were distinct and that their goals were occasionally different, he absolutely insisted that the long Christian tradition of natural theology was an essential underpinning of faith, and that man had to start with his experience of the physical world in order to transcend it. Given this assumption, buttressed by a thorough command of Aristotelian philosophy, St. Thomas insisted blatantly that there could be no conflict between reason and faith, or science and revelation.

Any apparent conflict had to be the result of a misinterpretation of either reason or revelation. It required the reassessment of both positions to discover where the error lay, though generally Thomas presumed the legitimacy of theological dogma and expected to find the errors in the philosophical tradition.

Had Thomas remained at Paris throughout the middle of his career, his moderating influence might have headed off the events that led to the condemnations in 1277. But he was gone from 1259 to 1269, and during this time the rift between conservative theologians and Latin Averroists became impossible to breach.

During the early 1260s Siger of Brabant lectured on natural philosophy, adopting Averroistic interpretations of Aristotle. In particular, he followed Averroes into a doctrine particularly abhorrent to the theologians, that of "monopsychism." According to this doctrine—which owes much to the astrological thought of Abu Ma'shar, among others—the intellectual aspect of man's soul is not merely created by God and placed in each individual, it is a part of God's intellect, temporarily implanted in man and returned to God on the death of the individual. How Averroes got to this point need not bother us here; except that it was through a detailed philosophical argument adapted from Aristotle's *De Anima* and with astrological overtones. As Gordon Leff pointed out, "For Christian Theologians this was to deny man any independent spiritual property that survived the destruction of his body,"[22] and it seemingly focused the nascent hostilities between the theologians and natural philosophers.

In a series of lectures delivered in 1267, St. Bonaventure outlined and began an attack on a series of errors being taught by Siger and his followers.[23] First and foremost of these was the doctrine that there was only one intellect shared by all men, so that individual immortality was impossible; but also included were the doctrines that the world is eternal, that God could not create out of nothing, and that in the physical world a complete determinism holds.

This attack on philosophical errors seemed to launch a whole conservative movement, exemplified by Giles of Rome's distribution of his *Errors of the Philosophers Aristotle, Averroes, Avicenna, Al Gazel, Al Kindi and Maimonedes*—and culminating in Tempier's condemnation of 219 theses in 1277. The key issue in these condemnations is that the "errors" are intentionally not philosophically refuted, as Thomas and the moderates would have wished. They are simply stated and shown (often only implicitly) to contradict a basic theological dogma or scriptural passage. Two examples from Giles's list of fourteen errors of Aristotle will serve to represent the whole class.

> (13.) Again, because he wished to proceed by way of nature, he believed, as is clear from book IV of the *Physics*, that since two bodies cannot *naturally* be in the same place, it was so essential for dimensions to resist dimensions, that it is impossible for dimensions to continue to exist and yet not resist dimensions.
> Because of this it follows that God could not make two bodies exist in the same place.[24]

Since God's *absolute* omnipotence must be held by faith, no limits on his will and activities can be set. It must be held that God, if he wished, could create two bodies in one place. Clearly, *any* statement from natural philosophy could in principle be undermined in this way.

> (14.) Again, because an intelligence could not have the potency for motion without being actually in motion since intelligences are placed in the best disposition if they move, he said that there were as many intelligences as there are [celestial] orbs, as is clear from book XII of the *Metaphysics*. "Divine scripture contradicts this, saying, 'Thousands of thousands ministered to him and ten thousand times a hundred thousand stood before him.' "[25]

In 1270, Stephen Tempier, Bishop of Paris, and a former member of the Parisian theology faculty, declared that "[the following] errors are condemned and excommunicated together with all who should knowingly teach or affirm them . . . (a) that the intellect of all men is one and the same; (b) that it is false to say that a man understands; (c) that men's will acts from necessity; (d) that all that happens in the world below is subject to the necessity of the heavenly bodies; (e) that the world is eternal; (f) that there was never a first man (that is, there was no act of creation); (g) that man's soul, which is his form, dies with him; (h) that after a man's death the separated soul (that belonging to the separated intelligence as a distinct spiritual being) cannot suffer from corporeal fire; (i) that free will is passive, not active; (j) that God does not know individuals; (k) that God knows only himself; (l) that there is no divine providence regulating human actions; in that God cannot make immortal and incorruptible what is mortal and corruptible."[26]

This list contains the most fundamental propositions of later and longer ones.

Tempier's declaration had virtually no immediate impact on the teaching of the arts faculty (though Siger apparently backed off on his monopsychism in response to a *philosophical* rebuttal by St. Thomas in 1270); so he appealed to the Pope (a former arts master at Paris). John XXI ordered Tempier to make a study of errors being propagated; the Bishop responded on March 7, 1277 with his condemnation of 219 propositions. Though he probably exceeded his authority from the Pope, he subsequently received Papal support when Siger of Brabant was called to the Papal Court and incarcerated until his death in 1284.

THE CONDEMNATIONS OF 1277: SCIENTIFIC AND THEOLOGICAL CONSEQUENCES

The consequences of Tempier's condemnations—or better yet, of the forces which gave rise to them—on medieval science and religion, and their interaction were profound, but so varied and complex that they cannot be followed here in all of their ramifications. We focus on just one specific line of development that had major long-term theological and scientific implications. This line of development moved through one of the compilers of Tempier's theses, Henry of Ghent, to William of Ockham, probably the greatest and most controversial theologian-philosopher of the fourteenth century.

Like Plato in the fourth century B.C., Ockham reshaped virtually all of theology *and* science by following up what was initially an antagonism to the reigning view of the nature of scientific knowledge. But whereas Plato turned science toward a search for purposes or final causes and toward a demand for absolute logical necessity in its conclusions, and in the *Timaeus* made science and theology inseparable, Ockham did precisely the opposite. He insisted on the radical dissociation of theology and science; he turned the sciences away from concerns with final causes; and he insisted that all knowledge of the physical world is contingent rather than necessary. Though his intent was almost certainly to enrich theology and protect it from philosophical attack, the result of his train of thought was equally certain to constrict the domain of Christian theology and to encourage a vast flowering of scientific thought. In C. Warren Hollister's graphic terms, "The Ockhamist position . . . blazed two paths into the future: Pietism uninhibited by reason and science uninhibited by revelation."[27]

Ockhamist thought begins with two outgrowths, or characteristics, of the conservative theology symbolized by the condemnations of 1277. In their zeal to undermine the authority of the natural philosophers, the Parisian conservative theologians raised to new prominence two ideas that had been present but subordinated in the Christian tradition. First was the absolute omnipotence of God and his total freedom from any constraints. Second was the consequent indeterminism of the physical world subject to God's will.

These principles were formulated by Ockham in the following three propositions upon which all of his theology and philosophy was based: (1) "All things are possible for God save such as involve a contradiction"; (2) "God can cause, produce, and conserve every reality, be it a substance or an accident, apart from any other reality"; and consequently, (3) "Everything that is real and different from God, is contingent to the core of its being."[28] Just one further proposition allowed Ockham to develop a new philosophy and theology:

> We are not allowed to affirm a statement to be true or to maintain that a certain thing exists, unless we are forced to do so either by its self-evidence or by revelation or by experience or by logical deduction from either a revealed truth or a proposition verified by observation.[29]

First let us briefly sketch the broadest implication of these principles for what we know as the scientific tradition. God's will is ultimately the first cause of all things; and because that will is in effect arbitrary and subject to none but logical constraints, it is fundamentally impossible to discover, in any specific way, the first causes of physical objects or events. Indeed, according to the second proposition above, it is strictly impossible to know whether any secondary cause is involved in establishing some phenomena either, since God could produce any object or event without a secondary (formal, efficient, or material) cause. Thus the old focus on causal analysis within studies of the natural world was undermined, or at least radically altered by the Ockhamists.

In order to understand what the Ockhamists (later called Nominalists) offered to replace the traditional scientific analysis we skip nearly three centuries in time to consider a statement from Galileo's analysis of falling bodies; then we look at how Galileo's statement can be linked to Nominalism.

In his treatise on *The Two New Sciences*, Galileo first correctly demonstrated that bodies fall at the surface of the earth in such a way that (ignoring the complications due to friction) the distance fallen is proportional to the square of the time of fall from rest. We express this relationship in the form $s = at^2$, where s is the distance fallen, t is the time of fall from rest, and a is a constant, which is the same for all bodies.

Galileo outlines the strategy used to reach this result early in the third dialogue of *Two New Sciences* in connection with a Nominalist criticism of Aristotelian causal analysis.

> The present does not seem to be the proper time to investigate the cause of the acceleration of natural motion, concerning which various opinions have been expressed by various philosophers, . . . Now, all of these fantasies, and others too, might be examined; *but it is not really worth while*. At present it is the purpose of our Author merely to investigate and to demonstrate some of the properties of accelerated motion (whatever the cause of this acceleration might be)—meaning thereby a motion, such that the momentum of its velocity goes on increasing after departure from rest, in simple proportionality to the time [i.e., $mv\alpha t$, where m is the mass of the body, v is its velocity, and t is the time] . . . and if we find [that] the properties [of accelerated motion] which will be demonstrated later are realized in freely falling and accelerated bodies, we may conclude that the assumed definition includes such a motion of falling bodies . . .[30]

Note that after he abandons the Aristotelian causal analysis, Galileo proposes two steps. First he will develop a mathematical description of uniformly accelerating bodies without asking whether such bodies exist or not. Second, he will experimentally or observationally determine whether there are real physical entities to which the theoretical or conceptual terms of his theory of uniformly accelerated bodies conform. These two steps correspond precisely to the two directions of investigation encouraged by Ockhamist thought. Indeed, as we shall see, the first step had already been made by the fourteenth-century figure, Nicole Oresme, who could hardly have been unknown to Galileo; and it was the second step in this case that represented Galileo's major contribution.

Ockham followed Aristotle in reserving the notion of science, strictly speaking, for a system of propositions *necessarily* derived from a set of first principles; but he diverged from Aristotle in two ways. First, he developed a set of new techniques (now embodied in modern propositional logic) which went beyond the old Aristotelian syllogisms in allowing necessary inferences to be made. Second, and more critical for our purposes, he denied that any first principles could have empirical content, since all experience is contingent rather than necessary. This meant that all true sciences had to be about nothing more than the concepts that the terms of the propositions signified. There could be strict and necessary sciences of Physics, Astronomy, Mathematics, and so on, as long as the scientist made no claim that the terms of the science referred to any objects that really existed outside of the mind of the thinker. One could demonstrate, for example, as Nicole Oresme did in a treatise entitled *On the Configurations of Qualities* (ca. 1348–1362) that *if* a body were to move with a "uniformly diform" motion in a straight line (i.e., in such a way that velocity increases uniformly in time, or $v\alpha t$) then the average speed could be written $v_{ave} = v_0 + 1/2\,(v_f - v_0)$, where v_0 is the initial velocity and v_f is the final velocity.[31] This proof plays a key role in Galileo's discussion of 1638, and it appears in a treatise that explicitly and emphatically insisted that such considerations are concerned with geometrical representations of hypothetical entities only and not with any empirical investigations.[32]

Bizarre as it may initially seem, the Ockhamist theoretical retreat from empiricism was singularly important in encouraging what we must acknowledge to be very important roots of theoretical—especially mathematical—physics, for in modern terms it made possible the birth of kinematics (the mathematical *description*

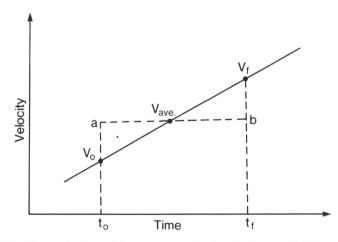

FIG. 17. Determination of the average velocity of a body exhibiting uniformly diform motion. For uniformly diform motion $V_{ave} = V_0 + \frac{1}{2}(V_f - V_0)$, since the distance traveled over time $t_f - t_0$ at $V_{ave} =$ area $t_0 ABt_f$ which in turn is equal to area $t_0 V_0 V_f t_f$.

of motion) separate from the discussion of the causes of motion.

But this does not constitute the only major thrust of Ockhamist thought. Ockham's statement regarding sufficient reasons for making an affirmation refers strongly to experience and observation. Though he insisted upon the contingency of experience and on its inability to establish a *science* in the older Platonic and Aristotelian sense, Ockham did not, therefore, believe that experience and observation were unimportant or incapable of providing the grounds for the most certain knowledge we can obtain about the world. In fact, what Ockham managed to do was to disengage the idea of what *we* would call "scientific" certainty from the older notion of philosophical necessity. While not all of the significant names in the growth of modern science would admit this distinction, many of the most important, including Galileo and Newton, would absolutely insist upon it.

Ockham again began from a strictly Aristotelian position before he diverged along a path that was to be tremendously liberating for the later scientific tradition. Aristotle's method held that all knowledge begins from experiences of primary substances or individual objects (see chap. 4). Of these individual objects, Ockham would admit that we have an absolute and intuitive certainty. Nothing can dissuade us of our awareness of their existence; so they may form the foundation of certain knowledge. But because of God's absolute omnipotence, he could in principle implant in us an intuitive certainty of the existence of something that does not exist. For this reason, though intuitive knowledge may be certain, it cannot be necessary. To many men this is a terrifying notion, for it admits that our most certain knowledge may in some sense be wrong. But Ockham insisted that it should hold no terror for us because, in terms that were introduced in chapter 1, what is relevant to our notion of scientific knowledge is our experiences (i.e., the phenomena). We cannot be wrong about our *experience* of the world, though we cannot infer from this experience the character of some "real" world presumed to underlie it. The thrust of Ockham's argument is thus to divorce physics from metaphysics.

We can have a certainty about the experienced world that, if we take proper precautions, will never be in error; this is what is important for Ockham, and for any claims made by all sophisticated subsequent supporters of experimental or empirical science. Thus Ockhamist, or Nominalist, thought provided important encouragement for the empiricist tradition of thought, which had other and equally important sources in the less radical Neoplatonic Oxford tradition of Grosseteste and Roger Bacon. A real alternative to the theoretical and purely hypothetical approach of men like Oresme was the direct recording and systematizing of our experiences of the natural world. This experimental approach had to renounce the old search for necessary causes—which could never be found, even if they existed in God's ineffable intellect—and had to seek merely to discover the constant correlations between experienced events; but it could provide the only foundation for certain and stable knowledge of the phenomenal world.

One must realize that just as few medieval thinkers became pure Aristotelians or pure Platonists, few if any became pure Ockhamists. Yet Ockham and his followers made available for synthesis into a wide variety of new systems of thought, elements that became more and more pervasive over time, and form the foundations of what we may call modern attitudes toward the nature and sources of scientific knowledge.

Ockham's response to the thirteenth-century controversy over the relations between faith and reason, theology and philosophy, or religion and science signalled a transformation of the scientific tradition; it also had major repercussions for the development of late medieval Christianity. For many it cut off the domain of faith from the supports it had garnered from traditional natural theology, and, perhaps more importantly, from the tradition of moral philosophy. Indeed, it undermined the whole theological tradition and placed a new emphasis on the individual Christian's direct intuitive confrontation (1) with God's revelation (the scriptures), and (2) with God's will (which might supercede his word). In this sense it provided a critical underpinning for later reformation theology.

Consider just two examples that illustrate the radical implications of Ockhamist theology. God's absolute freedom of will implied that he can will anything that does not imply a contradiction. He might, therefore, will that a man do anything, including even hate God. Now the first ethical responsibility of man is to do the will of God; thus it might be ethically good to hate God if he wills it.[33] This conclusion emphasizes that no act can be good or bad except in direct relation to God's will, and even such acts as murder and adultery can become justified when God reveals that they are approved. Thus there can be no "rational" support for Christian ethics.

By the same token, God's absolute freedom implies that he can choose to reward where there is no merit. Man's efforts to remain or become virtuous are thus of no real effect in leading to salvation, and God's grace is granted or withheld arbitrarily. Such ideas provide the cornerstone for a Protestant theology, which undermines the Catholic tradition of Christian ethics in favor of one of individual revelation.

In the mid-sixteenth century the Council of Trent was forced to reinstitute St. Thomas's theory regarding the relationship between faith and reason as the official position of the Catholic Church because church theologians could not cope with the religious anarchy attendant on decoupling Christian dogma from the traditions of natural and rational theology.

SUMMARY FOR CHAPTERS FIVE AND SIX

It makes no historical sense to try to understand the fundamental transformations of Christian thought from its origins in the Greco-Roman world through the so-called Dark Ages and into the High Middle Ages without reference to the classical scientific systems of Plato, Aristotle, and their Islamic commentators. By the same token, there is no way to understand the transformation of Hellenic and Hellenistic science with its overt metaphysical and theological emphasis into modern science with its overt divorce from theology and metaphysics, without reference to changes in Christian theology.

In its early formative years, Christianity came to have a vastly more favorable outlook on the physical universe than the other Eastern mystery cults with which it competed; and as a consequence Christian religion and scientific systems became intertwined in such a way as to make natural theology, grounded in the theory that God could be understood through the study of the world, a central feature of early Christian thought. Though some theologians, especially among the early Latin fathers, feared an overemphasis on reason and nature, Basil, Augustine, and others like Boethius who tended to dominate Christian education, insisted that the liberal arts, and chief among them logic, mathematics, astronomy and natural history, provided an important prerequisite for effectively understanding and advocating Christian doctrines. Thus, with the decline of the Roman Empire and decay of classical culture, at least some emphasis on preserving scientific knowledge was maintained in Western Christian monasticism.

With the economic revival of Europe in the tenth and eleventh centuries, and with the attendant expansion of Christendom into Spain and Sicily, which had been dominated by Islam, there was a tremendous revitalization of European learning, which depended heavily upon the absorption of Islamic knowledge with its orientation toward the sciences and toward Aristotelianism in particular. As this knowledge spread and became absorbed into the institution of the medieval university, secular learning, embodied in the newly emergent arts faculty and closely linked with Aristotelian ideas, was seen as increasingly threatening by many theologians. A renewed, but transformed, sense of conflict between the dogmas of Christian faith, grounded in revelation, and the doctrines of natural science, grounded in reason and experience, became a dominant feature of thirteenth-century intellectual life. This trend culminated in the condemnation of 219 (largely Aristotelian and Averroistic) theses by the Bishop of Paris in 1277.

Among the potential resolutions of the conflict, two with long-term importance emerged. Thomas Aquinas developed a scheme of reconciliation that denied the reality of any apparent conflicts, and reasserted the centrality of natural and rational theologies for the Christian tradition. Though this resolution ultimately became the foundation of subsequent Catholic theology, it was at least temporarily shoved into the background by the alternative, most dramatically posed by William of Ockham.

Ockham argued for the radical separation of theology and philosophy, and his critical thought led to major reformulations of both fields. In particular, Ockham fundamentally undermined the traditional claims of natural philosophers to be able to gain knowledge of the necessary causal relationships between phenomena, and produced a new theory of science that encouraged the growth of mathematical theory and of empirical investigations alike. At the same time the implications of his renunciation of traditional rational and natural theology involved a fundamental emphasis on faith and scripture that had a crucial bearing on Reformation theology. Both modern science and modern Protestant theology thus find an important stimulus in the outgrowth of thirteenth-century conflicts between science and religion, or more accurately, between theologians and natural philosophers.

The events leading up to and following from the condemnations of 1277 provide a marvelous example of how conflict between specialist groups in a culture can provide the mechanism for dynamic changes in both specialties.

The Rise of a Manipulative Mentality and the Democratization of Science, About 1370–1580

7

No intellectual historian of the Middle Ages could afford to ignore the relationships between science and theology; but perfectly respectable histories of medieval economic life and political life could ignore them entirely. Only within the universities and in connection with the practice of medicine did scientific ideas, attitudes, and activities have a major bearing on medieval life. During the period of the Renaissance and Reformation (ca. 1400–1600), however, a remarkable change took place. By the end of that period it was widely believed that mathematical reasoning and/or experimental studies (i.e., scientific techniques) provided the foundation for all technological innovation, for all forms of artistic expression, for legal systems, for commercial and economic activity, and for ethical and political understanding and progress. Among most intellectuals, by the early seventeenth century, theology alone was thought to be outside the proper domain of the sciences; and even on this point there was much disagreement. Western intellectual life had been scientized to an extent as great as that of late Hellenic Athens. But there was at least one great difference, which had to do with an understanding of the relationship between theory and practice, or between contemplation and activity in the world.

Aristotle had insisted that the subject matters of the most dignified "theoretical" sciences were beyond human capacity to alter in any way, and that knowledge *not* aimed at action is the most blessed and divine of human acquisitions. Early church emphasis on natural knowledge as a kind of revelation of immutable divine laws had maintained the Hellenic attitude toward the passive and receptive contemplation of the universe. But a new Renaissance attitude explicitly denounced this position, as we can see from one of its most eloquent expressions in *The Expulsion of the Triumphant Beast*, written in 1585 by Giordano Bruno:

> The Gods have given man intelligence and hands, and have made him in their image, endowing him with a capacity superior to other animals. This capacity consists not only in the power to work in accordance with nature and the usual course of things, but beyond that and outside her laws, to the end that by fashioning, or having the power to fashion, other natures, other orders by the means of his intelligence, with that freedom without which his resemblance to the deity would not exist, he might in the end make himself god of the earth. That faculty, to be sure, when it is unemployed, will turn into something frustrate and vain, as useless as an eye which does not see or a hand which does not grasp. For which reason, providence has decreed that man should be occupied in action by the hands and in contemplation by the intellect, but in such a way that he may not contemplate without action or work without contemplation. Thus, in the Golden Age [i.e., classical Athens], men, through idleness, were worth not much more than dumb beasts still are today, and were perhaps more stupid than many of them. But, when difficulties beset them or necessities reappeared, then through emulation of the actions of God and under the direction of spiritual impulses, they sharpened their wits, invented industries, and discovered arts. And always, from day to day, by force of necessity, from the depths of the human mind rose new and wonderful inventions. By this means, separating themselves more and more from their animal natures by their busy and zealous employments, they climbed nearer to the divine being.[1]

The major thrust of this and the next two chapters is to explore how the new attitude toward the relationship between thought and action came to prominence, how it fostered a linkage of the scientific tradition with the new manipulative mentality, and how this linkage had ramifications throughout Europe, both among traditional intellectuals and among the classes of artists, artisans, and agriculturalists whose economic activities were transforming Western society.

One special warning should be kept in mind throughout this chapter. Renaissance claims regarding the necessity of theoretical foundations for practical activities were made for a variety of motives. In some cases, for example, they were certainly merely expressed to lend dignity and status to activities that did not traditionally have them. In such cases, and often even where there is no question of the sincere belief of the speaker and writer, careful historical investigation shows that claimed connections are illusory. The historical inaccuracy of such beliefs does not, however, have much bearing on the significance of the beliefs themselves. Even if Renaissance figures were totally and always wrong (which they were not) in viewing mathematical theories and systematic experimentation as the source of past and present technological and intellectual progress, the fact that they believed them to be, and that they inculcated that belief in succeeding generations was of immense importance. The inheritors of Bruno's, and Bacon's, and Agricola's, and Alberti's, and Andreae's, and Vesalius's, and Vive's attitudes insisted upon seeking practical knowledge through scientific methods. For better or for worse their expectations were met in the long run, regardless of whether or not they were well founded at the outset.

The probability that sixteenth-century beliefs in the efficacy of applied science were often more prophetic than reflective of historical circumstances suggests that the relation between science and scientism in the early modern world is almost the converse of that in Hellenic Athens. In the ancient world, pre-Socratic natural philosophizing predated (and gave rise to) the scientistic visions of the Sophists, who hoped to extend scientific ways of knowing to wider ranges of topics. But in the Renaissance, widespread beliefs in the applicability of scientific methods to all problems facing humanity preceded rather than followed a dramatic resurgence of natural philosophizing. Indeed, it seems most correct to suggest that Renaissance scientism was a significant factor in generating enthusiasm and support for the activities we associate with modern science, activities that became widespread only during the seventeenth century after an extended period of scientistic preparation.

This comment may seem inconsistent with my claim in chapter 6 that one of the significant thrusts of Ockhamist philosophy and Oxford Neoplatonism alike in the fourteenth century was to produce a vigorous development of mathematical and empirical science that shows many modern tendencies—especially in connection with separating science from metaphysics. In fact, the relative importance of medieval Nominalist science and new Renaissance trends is hotly debated among historians of science. One cannot deny that in many ways the *content* of much Renaissance and early modern science shows substantial continuity with that of the fourteenth-century. We have already suggested this by relating Galileo's approach to that of Nicole Oresme. On the other hand, Ockhamist and fourteenth-century Oxford science was confined to a relatively small number of men working within the universities; and after a brief period of vigor in the fourteenth century, it became a relatively minor movement whose importance was only revived in connection with the humanistic trends of the Renaissance. When it was revived it found a vastly different institutional place, often outside of the older university setting, and this changed context was of immense significance.

For the sources of Renaissance scientism and the pressures to move natural philosophy beyond its old place in the quadrivium of the arts faculty we must look backward to the recovery and/or reinterpretation of the scientific knowledge and attitudes of antiquity. This recovery and reinterpretation was intensified in the humanistic movement centered in northern Italy during the fifteenth century. So it is to the context and content of Italian humanism that we must first turn.

EARLY HUMANIST OPPOSITION TO SCIENTIFIC THOUGHT

Early Italian humanism was explicitly and bitterly hostile to science as it was understood in the fourteenth century. We must understand the character of that initial opposition before we can see how it could gradually be transformed into support for a revised vision of science.

In some sense, humanism was as much a response to the issues symbolized by the condemnations of 1277 as the philosophy of William of Ockham. This can best be understood in connection with the writings of Francesco Petrarca (Petrarch, 1304–1374) who was admired by virtually all later humanists. Petrarch, like Ockham, was violently antagonistic to the Latin Averroeists. He even began writing a work, "against that ranting dog, Averroes, who, excited by hellish wrath, abused and blasphemed the holy name of Christ and the Catholic faith."[2] Moreover, Petrarch was at one with the conservative theologians in his belief that natural philosophy was incapable of penetrating God's essence.

> There are fools who seek to understand the secrets of nature and the far more difficult secrets of God, with supercilious pride, instead of accepting them in humble faith. They cannot approach them, let alone reach them. Those fools imagine they can grasp the heavens with their hands. Moreover, they are content with their erroneous opinion and actually imagine to have grasped the truth. Not even the telling words addressed by the Apostle to the Romans are able to deflect them from

their lunacy: 'Who knows the secrets of God? Who is a party to his counsels?'[3]

But Petrarch's partial agreement with the theologians should not mislead us too much. He was not a theologian. His training was in law. And as a student of law he was vastly more concerned with the subjects of the old trivium (especially rhetoric and grammar) than with the subjects of the old quadrivium. For Petrarch, rhetoric and classical literature (especially history), rather than classical *science*, should dominate education. Toward the end of his life he proposed a reform in education that would elevate the *studia humanitatus*—that is, the study of the moral sciences as they were represented in the works of the great masters of antiquity, especially Cicero and the Plato of the early Socratic dialogues—to centrality. Thus, Petrarch provided both the early central thrust and the name for the humanistic movement.

At least three features of Petrarchian humanism are important to bear in mind, for though not all later humanists maintained them, they were widely pervasive. First: humanism involved an initial new emphasis on the works of *pagan* antiquity—especially of Roman orators and historians like Cicero, Livy, Tacitus, and Plutarch; and this emphasis focused almost as much on rhetorical style as on content. But humanism was neither anti-Christian nor exclusively concerned with Roman exemplars. Petrarch, for instance, insisted on the first, writing:

> Christ is my God; Cicero is the prince of the language I use. I grant you that these ideas are widely separated, but I deny they are in conflict with each other. Christ is the Word, and the Virtue, and the Wisdom of God the Father. Cicero has written much on the speech of men, on the virtues of men, and on the wisdom of men—statements that are true and therefore surely acceptable to the God of Truth.[4]

Given its special concern with style and issues of ethics and morality, a second crucial aspect of humanism becomes understandable. Petrarchian humanism was particularly partial to Plato, among philosophers; it was soon realized that the Socratic dialogues of Plato were among the most eloquent examples of Greek rhetorical technique and that their central concern was moral and ethical. It was neither the idealist metaphysics nor the emphasis on mathematics that initially drew humanists to Plato, as it had drawn some of the early church fathers; but in the long run Platonic cosmology and mathematicism did become important.

The third central feature of humanism is perhaps of greatest importance for our present concerns. Rhetoric, including the orations of Cicero and the historical writings of Plutarch, were from a radically different ancient educational tradition than the natural philosophical works of Aristotle—a tradition which, like the scribal education of the ancient Near East, was self-consciously aimed at action in civic life rather than contemplation of eternal truths. And this distinction was not lost on Petrarch and his followers. Cicero had written,

> There is nothing more pleasing to God who governs the world than men united by social bonds. . . . There is a place reserved in heaven where all those who have labored to preserve, augment, and assist the fatherland, can enjoy eternal blessedness.[5]

Petrarch embraced this notion and insisted that *ceterorum hominum charitas*—love for one's neighbors, or Christian charity— was the most fundamental motive for humanistic studies. We should keep this early humanistic mandate in mind later, when Francis Bacon, the greatest early modern spokesman for scientism, insists that, "It was from lust of power that the angels fell, and from lust of knowledge that man fell, but of *charity* there can be no excess, neither did angel or man ever come in danger by it."[6] Similarly, we should keep it in mind as Bacon redefines the character and aims of natural philosophy.

> If there is anyone on whose ear my frequent praise of practical activities has a harsh and unpleasing sound because he is wholly devoted to contemplative philosophy, let me assure him that he is the enemy of his own desires. In natural philosophy practical results are not only the means to improve human well-being, they are also the guarantee of truth. There is a true rule in religion that a man must show his faith by his *works*. The same rule holds good in natural philosophy. Science, too, must be known by its works. It is by the witness of *works* rather than by logic or even by observation that truth is revealed and established. It follows from this that the improvement of man's lot and man's mind are one and the same thing.[7]

Just as the humanistic rhetorician is to be judged not by the formal logic of his argument, but by his ability to persuade an audience to do good, so too the humanistically oriented natural philosopher is to be judged, not by the logical necessity of his conclusions, but by his ability to command nature in the service of mankind.

The step from Petrarch to Bacon is not an easy one; for early humanism had virtually no sympathy for natural philosophy at all. Petrarch expressed this initial opposition by attacking "people who know a great deal about wild beasts, about birds, and fish" and by arguing that such knowledge is pointless "if one has no interest in discovering the nature of man, whence man comes, where he goes, and why he is born."[8] In a more moderate and positive mood, Filippo Buonaccorsi reiterated this feeling and linked it to the initial emphasis on Roman orators.

> Many Egyptians and Greeks have shown a preference for contempla-
> tion and have written much about the beauty and wonders of Creation.
> But I praise and admire above all the Romans who neglected the
> excellences of individuals and the pleasures of the mind and, writing
> instead about laws and morals, were mindful of the needs of men.[9]

The opposition between the early humanists (most of whom had studied law) and the natural philosophers (many of whom turned to medicine) was expressed in a series of bitter debates over the relative merits of law and medicine. Coluccio Salutati, for instance, wrote *On the Nobility of Law and Medicine* in 1406, and Giovanni d'Arezzo wrote a dialogue, *De Medicinae et Legum Praestantia* (A Prelude to Law and Medicine), about 1440. In both of these tracts law is preferred over medicine for three related reasons. One goes back to a classical sense that the domain of morality and human responsibility is vastly more important than the domain of nature because in it lie the foundations of society and of civil life, which is the only life worthy of man. In this way, law allows man to live a life *above* that of nature. The second goes back to the early Christian sense that the domain of morality and human responsibility is vastly more important than the domain of nature because in it lies the possibility of salvation (i.e., we obey the laws of nature because we must; but we obey the moral law because we *will*). We thus deserve praise or blame only in our relations to moral laws and not to natural laws, and moral law becomes the source of salvation.

The third reason for preferring moral over natural knowledge is closely related to Ockhamist epistemology. Natural science, or certain knowledge of nature, is ultimately beyond human capabilities. But we can have full and complete knowledge of civil and canon law because, while divinely inspired, legal systems are *human* creations and therefore capable of being understood by human beings. In Salutati's words:

> They have their origin not in external things, but *in us*. . . . Thus we know them with such a certainty that they cannot escape us and that it is not necessary to seek them among external facts. . . . They contain man's natural reason which every sound intelligence can understand and discover by reflection and discussion.[10]

For many early humanists, then, not only was moral rather than natural law "the builder of cities and conqueror of nature"[11] as well as the source of salvation; it was knowable in a way that natural law was not.

Against these claims, the defenders of medicine during the early fifteenth century maintained an almost purely Aristotelian position. Niccoletto Vernia's writing on *Whether the Sciences of Civil and Canon Law are More Noble than Medicine* provides a good example. "Legislation aims at a certain happiness with regard to social life and with regard to communication in civic assemblies. But this is not the true happiness. . . . We can approach God only through contemplation . . ."[12] And Antonio de Feurraries points out that in the well-developed societies of ants and bees, both cooperation and altruism are more highly developed than in many human societies. The true measure of human dignity and nobility must therefore be in contemplation rather than in concerted action.[13]

This early radical separation between humanistic studies and interest in the natural world was soon undermined in connection with two related sets of circumstances. Of these, the earlier was purely secular and grew out of a trend toward civic aggrandizement, which brought together humanistic intellectuals and groups of artisans in northern Italy. Out of this contact grew strong traditions of art and engineering with a new theoretical or scientific orientation. These traditions in turn convinced humanists of the practical value of mathematical and natural knowledge—natural knowledge, not as it was embedded in Aristotelian systems, but as it was embodied in the experiences of those who worked with physical objects.

The second set of circumstances had a much greater religious component and was associated with (or at least symbolized by) the rediscovery and publication of the writings of Hermes Trismagistus by the humanist Marsilio Ficino in 1463. The Hermetic philosophy, which was presumed to be of great antiquity, linked together Mosaic law, Platonic cosmology, natural philosophy, astrology, alchemy, and other forms of natural magic in such a way as to make knowledge of nature the very foundation of Christian piety and worldly success.

THE SOCIAL AND INTELLECTUAL CONTEXT FOR RENAISSANCE ARTISTS AND ENGINEERS

Humanism developed in northern Italy during a period of rapid economic and social development. After a dramatic depopulation in the fourteenth century (due to recurrent waves of the plague), the northern Italian cities—Milan, Venice, Florence, Verona, Siena, Padua—grew rapidly during the early fifteenth century. What is perhaps even more important, their wealth grew even faster than their population. The northern Italian cities, unlike the urban centers of the rest of Europe at this time, were largely oriented toward international commerce and manufacture rather than toward local agricultural trade. Roughly a third of the Florentine population, for example, was involved in cloth trade and manufacture. And virtually all of the Venetian populace depended on oriental trade and the manufacture of fine glassware and ships. In these cities, the bankers, merchants, and large manufacturers became more important, both economically and in civil affairs, than the local landed aristocracy.

The northern Italian cities were also distinguished sharply from the rest of Europe because they were independent of any effective centralized secular authority. They preserved an old tradition that made their governmental and social structures more like that of the ancient Greek polis than like any contemporary cities or nations. This similarity to the old Greek city-state carried over to governmental structures—which ranged from despotic through oligarchic to republican—and to intense rivalries, which made local patriotism, intercity warfare, and rapidly shifting alliances central facts of Italian city-state life.

Humanistic scholarship served the ideological needs of this society and shaped its goals in critically important ways. The humanists' emphasis on the active life spread beyond an enthusiasm for law and politics into a celebration of economic life in general. Matteo Palmieri, for example, in *On Civil Life* wrote,

> He who puts all his industry and all his care into honest things which are worth knowing and the knowledge of which results in either a private or common advantage, deserves to be praised . . .

and

> He who increases his property through honest skills without causing damage to anyone, deserves to be praised . . .

conversely,

> People who despise what is useful and what can be achieved reasonably, deserve to be reproached and cannot be considered virtuous.[14]

In a marvelously apt myth reported by Leon Alberti, seven goddesses, representing harvests, hearths, clothing, houses, gods, protection, and safety, through the paternity of Labor and under the guidance of Politics, give rise to Life, Industry, Virtue, Victory, Plenty, Truth, and Joy; that is, all good things come through efforts to increase material goods and provide military might.[15]

If humanists provided an *ex post facto* rationale for private and often exploitive economic activity, however, they also provided guidance for how accumulated wealth should be used. In general, the humanistic historical awareness of the great ages of Greece and Rome led to an intense emphasis on the recovery of that greatness by the Italian city-states. Thus, public glory and splendor manifested in military might, great artistic and architectural productions, and impressive public ceremonies (which included both music and drama) became critically important aims. They were the concrete embodiment of civic virtue, providing a kind of immortality for the city in which they appeared, for the patrons who made them possible, and for the artists and artisans who planned and produced them.

In the writings and attitudes of the painters, sculptors, architects, and military engineers who emerged in response to these public appetites for security, splendor, and spectacle, we find the first major linkage between humanistic and scientific attitudes and activities.

OPTICS AND MATHEMATICS IN THE SCHOOL OF BRUNELLESCHI AND ALBERTI

In my attempt to characterize the vital humanistic movement in early fifteenth-century Italy, I have ignored the continuing presence of a significant medieval tradition of Averroistic and Nominalist natural philosophy within the universities. Such a tradition did continue—especially at Padua—and one of its products played a particularly important role in the next phase of our story.

The science of optics had been extensively developed within the Oxford and nominalist traditions of late medieval natural philosophy, and the optical tradition involved both important mathematical and empirical components. The scientific study of optics, which began in some sense with Euclid's *Optics* and continued through the Islamic works of Alhazen, the works of Roger Bacon and John Peckham of Oxford, and the spectacular analysis of the rainbow by Theodoric of Freiberg, was taught at Padua by Biagio Pelaconi, who focused on the problem of how the appearance of objects is related to their positioning in space. Biagio's investigations were summarized in a work written around 1390 and entitled *Questiones Perspectivae*.

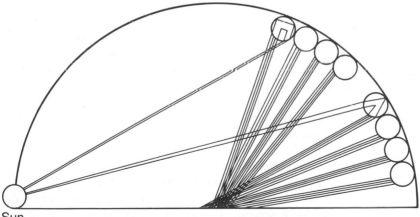

Sun

FIG. 18. Theodoric of Freiberg's explanation of the primary and secondary rainbows. Different colors are refracted differently entering into the water vapor droplets, then the rays undergo two total internal reflections to produce the primary bow and three total internal reflections to produce the secondary bow.

This work was brought to Florence by a young mathematics student named Paulo Toscanelli, whose major interests were in astronomy and astrology, but who was sufficiently fascinated by questions about optics and perspective that he wrote his own work on the subject.[16] There is evidence that when Toscanelli arrived in Florence there was already substantial interest in perspective, for a massive collection of texts on perspective was put together about 1400 by a Florentine humanist scholar who deposited the manuscripts in the Library of San Marco. These texts, which include the *Perspectiva* of John Peckham, appear to have belonged originally to Coluccio Salutati, but it is unclear how widely they were discussed and studied.[17]

Regardless of what had happened prior to Toscanelli's first return to Florence from Padua sometime before 1424, a dramatic change in the direction of Western art and architecture undeniably occurred as Toscanelli's scientific interest in perspective was joined to the practical artistic and architectural interests of Filippo Brunelleschi (1377—1446). It is equally certain that the linkage between scientific studies, art, and architecture symbolized by Brunelleschi's work had a substantial long-term impact on Western science and scientism as well. Giorgio de Santillana has even claimed that, "Brunelleschi around 1400 should be considered the most creative scientist as well as the most creative artist of his time."[18]

Filippo Brunelleschi was born into a wealthy Florentine family in 1377. His father was a government functionary, and his mother a member of the noble Spici family. After what seems to have been a very rudimentary formal education, Filippo was apprenticed to a goldsmith, as were the other great artist-sculptors of his time, Donatello, Ghiberti, and Ghirlandaio. In this setting, Brunelleschi joined with other young studio artists and began to study "motion, weights, and wheels, how they may be made to revolve and what set them in motion, and so produced with his own hand some excellent and very beautiful clocks."[19]

After losing a competition for producing the baptistry doors for the church of St. Giovanni, Brunelleschi went to Rome with Donatello to study architecture. Among other things during this period, he undertook a series of studies of ancient Roman buildings, using a tape and uniform measure to establish their exact dimensions. As de Santillana has pointed out, this, "may look simple enough, but is described as a great new idea. The old master builders had never come upon it. They did not write. They took their measures, we are told, with pieces of string, and left it at that. No collaboration of practical necessity with the wisdom of the learned monks who did advise them, no interplay of scholarly thought with technology had gotten beyond this stage for years."[20] At the same time, and probably during his frequent returns to Florence (which he made to replenish his funds by making clocks) Brunelleschi came into contact with the young Toscanelli and began studying perspective with him. Very rapidly, the artisan made a major experimental discovery related to optics, architecture, and perspective; developed the theory of linear perspective for painting; and taught his theory to Masaccio and probably to others.

Brunelleschi's first perspective device was described by his friend and contemporary Manetti. On a wooden surface about twenty-two inches long he painted the square of the Cathedral of Florence, seen from a point three feet inside the main door of the cathedral. The sky was filled in with burnished silver "so that the air and sky should be reflected in it as they are, and so the clouds are seen moving on that silver as they are born by the winds." At the point where the perpendicular of vision met the painting, Brunelleschi drilled a small hole which funneled out to the back of the wood. Opposite the painting, about an arm's length away, he mounted a mirror. Now, if one looked through the hole at the back of the picture,

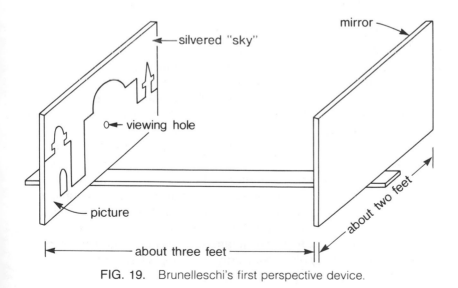

FIG. 19. Brunelleschi's first perspective device.

one saw the scene from its perspective point, "so that you thought you saw the proper truth and not an image."[21] Leon Alberti, Brunelleschi's close friend, reversed this procedure, creating a *camera obscura* in which light coming in through the hole projected the scene on oiled paper. But Brunelleschi's instrument was of significant architectural value, for he could see precisely what a proposed building would look like in its setting simply by painting a view of it in his device.

Within a few years of Brunelleschi's utilization of linear perspective, fascination with the technique had transformed Italian painting and sculpture, and the training of artists and had given rise to a new literature on painting. Leon Alberti first wrote out a description of the new techniques in *On Painting* which appeared in 1435, and Paolo Ucello, another Florentine contemporary, reputedly wrote a book on perspective about the same time. Ucello's obsession with its problems clearly appears in his drawings. Soon after, Piero della Francesca, who studied for ten years in Florence, wrote *De Prospectiva Pingendi* which introduced the so-called vanishing point method of perspective. Leonardo da Vinci introduced further complexities in his *Treatise on Painting*; and the German artist, Albrecht Dürer, who visited Italy several times, wrote on perspective and extended his mathematical studies to the proportions of human bodies in his *Treatise on Mensuration* (Nuremberg, 1525) and his *Four Books on Human Proportion* (Nuremberg, 1528).

George Sarton has quite correctly pointed out that the level of mathematical and optical knowledge required for linear perspective drawing is not very great.[22] But that does not justify his related implicit assumption that it is therefore unimportant, either as an element in the scientizing of other cultural specialties or as an event in the development of the scientific tradition itself. Alberti's claim in *On Painting* that "our instruction in which all the perfect absolute art of painting is explained will be easily understood by a geometrician, but one who is ignorant in geometry will not understand these or any other rules of painting"[23] is a bit overstated; but it is true that some familiarity with Euclid's *Elements* and *Optics* would have helped in following his discussion of the visual pyramid and its uses in accurately constructing the projections of horizonal figures on the vertical plane of a painting. It is even more true that Italian painters took Alberti's admonition to learn geometry[24] seriously as they experimented with a variety of increasingly complex systems of double and triple vanishing point perspective.

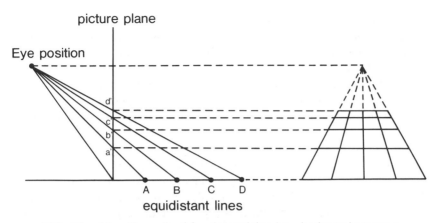

FIG. 20.   Alberti's method for determining how horizontal squares should be portrayed in a vertical painting.

It is often claimed that the interest in mathematics displayed by men like Brunelleschi, Alberti, and later Leonardo, derived from idealistic Neoplatonic metaphysics, and was closely linked with number mysticism; we will soon consider a tradition that *was* deeply informed by Neoplatonism. But it is important to recognize that the mathematical concerns of Alberti and other artists and artisans seldom, if ever, committed them to any particular philosophical stance.

In his consideration of optical rays, for example, Alberti intentionally sidesteps key philosophical issues to focus on practical ones. "Among the ancients there was no little dispute whether these rays came from the eye or the plane. This dispute is very difficult and is quite useless for us. It will not be considered."[25]

Far from being closely linked with Platonic idealism, artist-artisan mathematics was in fact usually linked with a strong emphasis on sensory experience. A Neoplatonic emphasis on mathematics was intended to draw the thinker away from mere experience to a consideration of the transcendent God or Good. The Renaissance artistic emphasis on mathematics was rather intended to help the artist enter into, and more precisely order, experience. Alberti made this point beautifully at the very beginning of *On Painting*:

> Mathematicians measure with their minds alone the forms of things separated from all matter. Since we wish the object to be seen we will use a more sensate wisdom.[26]

This focus on sensate wisdom in Alberti's mathematical discussion is clear from the fact that instead of talking about ideal lines and angles and surfaces, he explicitly discusses geometrical relations in terms of how and where to draw lines on a canvas and how to mark them off according to a measure derived from the length of a man's torso. Similarly, he describes the use of a grid between the painter and scene to be reproduced in order to properly place elements of a painting. But even more important is the linkage in such treatises between mathematical organizing principles and intensely empiricist emphases. Alberti, for example, constantly emphasizes that it is only from the close investigation of nature that one learns the proper measures of man.[27] He insists that the movements of the body and expressions of emotions be closely studied,[28] and he emphasizes that "All that I have said about the movements of animate and inanimate objects, I have *observed*."[29]

Some recent historians of science have argued that one of the important characteristics of the so-called scientific revolution was a change in the way men thought about mathematics from a Platonic (mystical and idealist) mode to an Archimedian (applied and empirically oriented) mode.[30] If this is true, this change gathers momentum in the writings on perspective of the early fifteenth century. But if this changed attitude had been limited to painting, it would have been relatively short-lived and unimportant. The new emphases on mathematics and empiricism became vastly more important in connection with architecture and engineering. And once more Filippo Brunelleschi and Leon Alberti stand out as among the earliest practitioners and literary spokesmen of the new movement.

The movement was effectively launched in connection with a single dramatic architectural feat, the construction of the dome of the Santa Maria de Fiore in Florence. Construction of the cathedral had begun at the very beginning of the fourteenth century, and part of the plan had involved a huge free-standing dome, spanning over 270 feet. If traditional methods had been used to build the dome, a huge timber scaffolding would have been built inside to support successive courses of stone until a keystone was laid in the center, after which the lower portions of the scaffolding would have been removed. But by about 1400, when work on the dome was due to begin, there was a severe timber shortage, brought on principally by deforestation to construct ships, and building such a timber scaffold was out of the question. Even had materials been available, it was not clear that such a huge timber structure could support its own weight, let alone that of the mammoth stone dome.

In 1407 a meeting of renowned master builders was held in Florence to determine how to complete the structure. No resolution of the problem was discovered, so an open competition was held in 1418. This competition was won by Brunelleschi, who proposed a radical new technique based on theoretical considerations rather than on traditional building methods.[31] The technical details of his double-shelled rib and sail method are beyond the scope of our considerations. What is vastly more important are the social changes inherent in Brunelleschi's work.

When Brunelleschi took over direction of the work on the cathedral he immediately ran into a problem when the conservatively minded masons and builders balked at working in his prescribed way. A bitter battle ensued as Brunelleschi fired workers wholesale for obstructionism. In response, their guild, the Guild of Masters of Stone and Wood, sought to have him jailed for failing to pay dues to them. Brunelleschi appealed to the powerful group of wealthy and civic-minded men who constituted the Works Committee for the Cathedral. This group supported him and jailed the guild leaders instead until they agreed to do things his way. Thus did Brunelleschi overcome the tradition of those he sneered at as "Grand masters of the trowel."[32]

Giorgio de Santillana has characterized the implications of this confrontation in a way that I cannot improve upon:

> Here we have, then, for the first time the Master Engineer of a new type, backed by the prestige of mathematics and of the 'recondite secrets of perspective' (Galileo's tongue-in-cheek description of his own achievements is certainly valid here), the man whose capacity is not supposed to depend only on long experience and trade secrets, but on strength of intellect and theoretical boldness, who derides and sidesteps the usual thinking-by-committee, who can speak his mind in the councils of the city and is granted patents for his engineering devices, such as that one of 1421 which describes in the somewhat casual Latinity of those circles a machine 'protrahendo et conducendo super muris Cupolae lapides, macignos et alia opportuna,' or the other, of the same year, for a river lighter equipped with cranes. He is, in fact, the first professional engineer as opposed to the old and tradition-bound figure of the 'master builder': He is the first man to be consulted by the Signoria as a professional military engineer and to design the fortifications of several towns: His work is the precondition of the first text-book, that of

Francesco di Giorgio Martini. But still and withall, he is acknowledged to the end of his life as the great designer and artist; not only that, but as the man who masters the philosophical implications of what he is doing. Donatello may be acquainted with the Latin classics; he is still considered a craftsman. Brunelleschi is not; he stands as an intellectual.[33]

Just as Alberti popularized and spread knowledge of Brunelleschi's integration of scientific and artistic activities in *On Painting*, he provided the literary counterpart of Brunelleschi's new theoretical approach to architecture and engineering in *De Re Aedificatoria* (*On Architecture*), which was written by 1452 and published in 1485. This treatise was the first printed work on architecture, predating even the first printed edition of Vitruvius' *Ten Books on Architecture*, the classical model on which it was based. And it had a major impact on the next generation of Italian architects and engineers.[34]

Unlike medieval treatises on architecture, which offered descriptions of precisely what had to be done in what order, but without any general principles that could help meet new problems, Alberti aimed at providing a theory of architecture grounded in knowledge of mathematics and physics. Erwin Panofsky writes that Alberti,

> basing himself on Vitruvius, but varying, expanding, and even correcting him in all directions, he derives his prescriptions from general principles such as practical purpose, convenience, order, symmetry, and optical appearance. He divides the tasks of the architect into different classes, which, taken together, form a coherent and comprehensive system from city planning to the construction of fireplaces, and he tries to corroborate his statements both by deductive, though naturally not always critical, reasoning and by historical evidence.[35]

THE EXTENSION OF SCIENTIFIC ATTITUDES AMONG RENAISSANCE ENGINEERS, ANATOMISTS, AND ARTISANS

Just as it is true that a low level of mathematical knowledge was necessary for perspective studies, it is also true that a limited mathematical sophistication and knowledge of "scientific" mechanics was evidenced by Alberti in *On Architecture*. But the book polemicized for increasing mathematical knowledge and knowledge of such topics as mechanics and optics, and sent budding young architects and engineers to recovering ancient scientific knowledge and developing their own new theories—especially in connection with ballistics and fortifications. Within a few years of Alberti's treatise Antoni Filarete suggested in his own architectural treatise that the works of Archimedes should be studied directly by all architects and engineers. Shortly thereafter the foundations of the modern sciences of mechanics and mathematics began to appear in the writings of military engineers like Nicolo Tartaglia and Guido Ubaldo.[36]

It may well be the case, as A. R. Hall has tried to show in his *Ballistics in the Seventeenth Century*,[37] that the theoretical works of these engineers had relatively little immediate impact on the practical activities they hoped to transform. But that fact is simply irrelevant to two dramatic aspects of Renaissance engineering. First, the cultural specialties of civil and military engineering, which had been all but totally separated from scientific attitudes and activities since the Hellenistic period, were infused with scientific attitudes, activities, and products. As a result, most engineers after the mid-sixteenth century admitted that they had to learn to use the knowledge produced by scientists (i.e., by the likes of Archimedes and Hero of Alexandria). Many of them so deeply imbibed the scientific attitudes that we discussed in chapter 1 that they became convinced that technological *advance* must presuppose *new* scientific knowledge; thus these men sought, like Tartaglia, to create a *New Science* (1537) of ballistics, or like Guido Ubaldo, to make new discoveries in such old sciences as *Mechanics* (1577).

The initial thrust of the Renaissance artistic and engineering union of theory with practice, especially as it was publicized by humanistically trained men like Alberti and Filarete, was to raise the status of painting and engineering to that of gentlemanly pursuits. Thus, Danielle Barbaro's introduction to his Italian translation of Vitruvius contains a marvelous passage on the status of various arts.

> The arts that serve, with dignity and greatness, the convenience and use of mortals—can be considered in two ways. First by how they are discussed and how, by ways of reasonings, they find the causes and rules of operation; second, by way of working on some external matter with manual dexterity. But the following criterion enables us to know the more worthy arts; all those which require the sciences of numbering, geometry and mathematics [i.e., navigation, the military art, the art of building, medicine, painting, and sculpture]—all such arts have a quality of greatness about them. The rest . . . [are] sordid and baser.[38]

One can see a second fine example of this tendency in George Agricola's *De Re Mettalica* (1556), an extremely widely used treatise on mining engineering. Agricola tries to eradicate the stigma of the older notion that mining was a "servile, sordid toil, unworthy of a free man or an honest gentleman."[39] In order to do this he insists that the technical aspects of mining, in fact, presuppose a mastery of several fields of knowledge, including minerology, mathematics, philosophy ("that he may discern the nature of subterranean things"), medicine, and even law.[40]

But Agricola takes a second critical step. He also attacks the ancient aristocratic linkage between slavery and manual activities.

> Certainly, if mining is a shameful and discreditable employment for a gentleman because slaves once worked mines, then agriculture also will not be a very creditable employment, because slaves once cultivated fields; nor will architecture be considered honest because some slaves have always been found skillful in that profession; nor medicine, because not a few doctors have been slaves. Nor will any other worthy craft, because men captured by force of arms have practiced it.[41]

The first of Agricola's arguments essentially accepts the older Aristotelian notion that knowledge is superior to practice and tries to dignify practice by linking it to knowledge. But the second line of reasoning opens up a new direction that calls for comment. Agricola shows a clear awareness that classical arguments in favor of theory rather than practice were grounded in class biases that valued leisure over practical economic activity. But these older class biases were becoming irrelevant in a Renaissance society in which power was moving into the hands of bankers, merchants, and industrialists whose wealth and position derived from economic activity.

A sensitivity to the shifting values of Renaissance society certainly appeared among early sixteenth-century humanists, who were only marginally aware of the engineering tradition. Thus, for example, Thomas More cried out in his *Utopia* (1516) against those "noblemen among you that are as idle as drones, that subsist on other men's labor, . . . [and] who grow feeble with ease and are softened with their effeminate manner of life."[42] He insisted that among Utopians all men learn a trade, and that the chief business of the government "is to take care that no man lives idle."[43]

Nothing like More's radical communistic alternative to traditionally class-dominated society was remotely possible in Renaissance Europe. But authors who concerned themselves with aspects of technology produced a growing stream of polemics designed to promote the dignity of the manual arts, independent of their intellectual pedigree and based solely on their utility. And this raised claim of dignity clearly attached itself to the artisan as well as his art. The Italian translator of Guido Ubaldo's *Mechanics* (1581) provides a particularly nice example. He defends mechanics as a most honorable term ". . . appropriate to a man of high position who knows how with his *hands* and *heart* to carry out marvelous works of singular utility,"[44] and he specifically attacks Plutarch's claims (see chap. 4) that Archimedes "disparaged mechanics as base and vile and material and did not deign to write of it."[45] He points out that, in fact, Archimedes not only performed mechanical tasks but also wrote many works on them, and that Archimedes' great reputation in antiquity came more from his military inventions than from his theoretical works.

The upshot of this tradition of writing was almost to reverse the traditional emphasis on the relationship between theory and practice. If practical activity had initially gained stature by establishing ties to "pure" knowledge, more and more, as time went on, knowledge was seen as worthwhile only when it could be directly linked to manual operations. This reversal of values, moreover, appeared not only in the writings of artisans who insisted on their own dignity, but in the writings of many sixteenth-century humanists who sought to preserve the initial humanistic emphasis on service to mankind against a growing tendency among academic humanists to retreat into a kind of uncritical adoration of classical authors, and an exclusive emphasis on rhetorical style and bookish learning.

It is worth pointing out that the newly perceived two-way interdependence of theory and practice had immediate ramifications for the development of science; just as the confrontation of artisan and scientific subcultures produced a transformation of artisan values and attitudes, it produced substantial changes in the attitudes of the scientific subculture as well. Paolo Rossi has argued, for example, that the growing sixteenth-century demand for "a knowledge in which the observation of phenomena, attention to operations, and empirical research would have a superior status vis-à-vis rhetorical evasions, verbal accommodations, logical subtleties, and a priori constructions," a knowledge sharply critical of merely aristocratic and bookish attitudes, was nowhere more powerfully stated than in Andreas Vesalius' *Fabric of the Human Body* (1543), perhaps the greatest text in the renewal of anatomical studies.

Vesalius, who had studied under Gunther von Andernach, the humanistic translator of Galen's *On Anatomical Procedures*, saw that the level of anatomical knowledge had dramatically declined since antiquity; he placed the primary responsibility for that decline on the classical aristocratic disdain for manual activity.

> After the barbarian invasions all the sciences, which before had gloriously flourished and been rightly practiced, went to rack and ruin. At that time, and first of all in Italy, the fashionable doctors began to despise the work of the hand, in imitation of the ancient Romans. . . . In this way, unfortunately, they estranged themselves from the most important and most ancient branch of the art of medicine, that which is based above all on the investigation of nature. . . . When the whole procedure of the manual operation was handed over to barbers, the doctors not only soon lost the true knowledge of the viscera, but very soon, practical anatomy also came to an end.[46]

To reverse this trend Vesalius insisted that anatomy be learned not merely from the texts of Galen and others, but through hands-on experience in dissection. In this way anatomical science was reborn and rapidly expanded in the late sixteenth century.

SUMMARY

After an initial intellectualizing or "scientizing" of technical activities, beginning with painting and architecture in the early fifteenth century, and navigation in the sixteenth century,[47] the linkage between theory and practice in the arts had a major twofold impact. It emphasized the scientific elements of practical arts, and thereby vastly expanded the potential audience for scientific knowledge just at a time when printing came into existence to make the dissemination of that knowledge possible beyond the confines of the universities. The publication and republication of numerous books on anatomy, painting, architecture, military engineering, machine construction, navigation, and commercial arithmetic, as well as of such classic texts as Euclid's *Elements*, attests to the demand for knowledge. And the content of such works indicates the emphasis placed on the scientific elements of practical arts during the Renaissance.[48]

One clear result of this expanded audience for science was to "democratize" intellectual life. Whereas the old universities served primarily clerics and lawyers whose backgrounds and aims linked them with an aristocratic elite, new institutions like Gresham's College, founded by the city of London and the Company of Mercers (cloth traders) in 1598 to serve the needs of the inhabitants of London, taught in English for those who did not understand Latin, and emphasized the applications of physic (medicine), geometry, and astronomy to trade, commerce, and navigation.

This democratization of intellectual life had a reciprocal influence on the scientific tradition from which it drew. As the applicability of science to practice became an increasingly widespread cultural assumption it was also argued that scientific knowledge not only could, but should, be enriched by the kind of direct experience of nature that artisans alone had embraced during the Middle Ages. First in writings about painting, like Alberti's and Leonardo's, then in those on mining and military engineering, like George Agricola's and Guido Ubaldo's, and on anatomy, like Vesalius', and even in the writings of literary humanists and educational theorists like Rabelais and Juan Louis Vives, one finds an insistence that knowledge must be sought directly from nature and in the results of operations on natural objects. Vives' sums up the thrust to dignify experimental over merely speculative or theoretical knowledge by insisting that scholars begin to "enter into the workshops and factories, asking questions of the artisans, and trying to be cognizant of the details of their work."[49] Furthermore, he contrasts the solid natural knowledge derived from experience against that of the so-called philosophers, who, "have constructed [nature] for themselves; that is to say, the nature of formalities, individualities, relations, of Platonic ideas and other monstrosities which cannot be understood even by those who have invented them."[50]

The speculative tradition of natural philosophy associated with Aristotle did not immediately disappear under attack from disaffected humanists and rising artisan-engineers. Instead, the two traditions developed in uneasy tension with one another throughout the early modern era; and the Aristotelian tradition remained dominant within the extremely important universities well into the seventeenth century. But the infusion of new practical concerns would become increasingly important over time.

Magic, Astronomy, and the Governance of Nature, Church, and State, About 1460–1633

8

While a secular antiphilosophical tradition that linked science to practice was emerging from painting and engineering, with new developments in mathematics, mechanics, and anatomy, a very different religious and philosophical tradition was also emerging that reinforced the growing links between theory and practice. This tradition emphasized the old sciences of astronomy and/or astrology, alchemy, medicine, and cosmology. It began in connection with the recovery and publication of a strange set of writings associated with the name of Hermes Trismagistus, who was thought (until early in the seventeenth century) to have been an Egyptian wise man who lived sometime around 2000 B.C.

HERMETICISM, NATURAL MAGIC, AND RENAISSANCE SCIENCE

In 1463, as Marsilio Ficino was translating the corpus of Platonic writings at Florence under the patronage of Cosimo de Medici, Cosimo's agents returned from Macedonia with a set of manuscripts purporting to be those of Hermes. Hermes was widely known by reputation, since he had been mentioned as a great wise man by Clement of Alexandria, St. Augustine, and Lactantius. In addition, the Hermetic corpus had been referred to in connection with the origins of Islamic astrology by Abu Ma'shar, and in connection with the writings of Roger Bacon. A very brief consideration showed Ficino that the Hermetic writings contained remarkable parallels both with the Mosaic account of Genesis and with the Platonic cosmology of the *Timaeus*; so he dropped the Plato translations and

230

rapidly produced a translation of the Hermetic writings under the title *Pimander*.

From our twentieth-century perspective it is unsurprising that the Hermetic writings showed similarities with Christian and Platonic ideas; for we know now that the writings originated in connection with a second- and third-century Gnostic movement centered in Alexandria, and that they drew heavily from both Christian literature and the writings of Plato and early Neoplatonists. But when the Hermetic writings appeared in 1463, they appeared as true marvels—as anticipations of central Christian and philosophical doctrines written a thousand years before Moses, and nearly fifteen hundred years before Plato. Their authenticity was presumed because of favorable references from the early church fathers. And their presumed antiquity linked with their supposed status as historical sources of Christian doctrine and Platonic philosophy ensured an audience for Hermes. This is attested by the existence of a large number of manuscript copies and of fifteen printed editions of the *Pimander* between 1471 and 1500.[1] In 1469 a second Hermetic work, *The Asclepius*, was also published; and the combined works were frequently reprinted throughout the sixteenth century. Politicians, poets, professors, and priests began to supplement Aristotle, Plato, Cicero, and the church fathers with Hermes as a source of stirring quotations; and the artists decorating the Cathedral of Siena made him the central figure of their grand mosaic.

The great question for us, of course, is what did the Hermetic writings contain that has a bearing on our concern with science, scientism, and society? To answer this question we must consider how the nature of man, the physical world, and man's relations with the physical world were portrayed in the *Pimander* and *Asclepius*.

Before I try to summarize what seem to be the most important features of Hermeticism, I should issue a caution to those who might want to consult the works themselves.[2] The fourteen books of "Hermetic" writings translated by Ficino were not all by one author. In fact, internal evidence suggests that several of the books may have been written by multiple authors. To make matters especially confusing, the different books contain mutually contradictory attitudes; so there is no single systematic Hermetic doctrine to be discovered in *Pimander*. That a relatively coherent Hermetic tradition emerged from these writings is largely due to the fact that Ficino and subsequent Christian humanists chose to emphasize certain aspects of the writings rather than others. The things they emphasized were certainly there, but they were accompanied by alternative points of view that were selectively ignored.

The most important unifying theme of Renaissance Hermeticism is powerfully presented in Book I of *Pimander* and in the *Asclepius*. In Book I of *Pimander*, the divine *Nous* (Pimander), appears to Hermes in a dream. Pimander turns into a blinding light, from which emerges a holy Word. Pimander explains: "That light is myself, *Nous*, thy God, . . . and the luminous Word issuing from the *Nous* is the Son of God."[3] Next, Hermes is shown that he contains within himself a comparable light, or *Nous*, with its attendant powers. Then Pimander begins to explain the origins of the sensible world. "The *Nous*-God, existing as life and light, brought forth a second *Nous*-Demiurge, who, being the God of fire and breath, fashioned the Governors, seven in number (i.e., the planets) who envelop with their circles the sensible world."[4]

The reader who has followed what was said in chapters 2 through 5 will recognize in this passage elements from the *Timaeus* and the related Alexandrian *Logos* doctrine as well as a Gnostic leaning in the assignments of the governance of the sensible world to the planets. But here Hermes begins to deviate from earlier traditional views as he describes the creation of man.

> Now the *Nous*, Father of all beings, being life and light, brought forth a Man, similar to himself, who he loved as his own child. For the Man was beautiful, reproducing the image of his Father: for it was indeed with his own form that God fell in love *and gave over to him all his works*. Now, when he saw the creation which the Demiurge had fashioned in the fire, the man wished also to produce a work, and permission to do this was given him by the Father. Having thus entered into the demiurgic sphere, in which he had full power, the Man saw the works of his brother [i.e., the Demiurge or Christ], and the Governors fell in love with him, *and each gave to him a part in their* [sic] *own rule*. Then, having learned their essence and having received participation in their nature, he wished to break through the periphery of the circles and to know the power of Him who reigns above the fire.[5]

Consequently, Man takes on a mortal body to live in nature, but he retains his divine powers not simply to *know* the world, but also to *share in its governance*. At this point, the Hermetic writings have certainly broken with the early Gnostic tradition and with the Platonic-Aristotelian-Christian tradition of contemplative rather than active knowledge of the world. In opposition to the older Gnostic tradition, Hermes portrays man not as subject to astral determinism, but as sharing with the stars in governing the world; and in contrast with the older Christian tradition, Hermes portrays man's likeness to God, not primarily in his ability to understand nature passively, but in his ability to emulate God's creativity.

These themes are reiterated with even greater clarity in the *Asclepius*.

> All descends from heaven, from the One who is the All . . . From the celestial bodies there are spread throughout the world continual effluvia, through the souls of all species and of all individuals from one end to the other of nature. Matter has been prepared by God to be the receptacle of all forms; and nature imprinting the forms by means of the four elements, prolongs up to heaven the series of beings.
>
> All species reproduce their individuals, whether demons, men, birds, animals, and so on. The individuals of the human race are diverse; having come down from on high where they had commerce with the race of demons [i.e., celestial deities] they contract links with all other species. That man is near to the Gods who, thanks to the spirit which relates him to the Gods, has united himself to them with a religion inspired by heaven.
>
> And so, O Ascelpius, man is a *magnum miraculum*, a being worthy of reverence and honour. For he goes into the nature of a god as though he were himself a god; he has familiarity with the race of demons, knowing that he is of the same origin; he despises that part of his nature which is only human for he has put his hope in the divinity of the other part.
>
> Man is united to the gods by what he has of the divine, his intellect . . . [but] seeing that man could not regulate all things unless he gave him a material envelope, He [God, the Father] gave him a body. Thus, man was formed from a double origin, *so that he could both admire and adore celestial things and take care of terrestrial things and govern them.*[6]

Once again, Hermes linked a transformed doctrine of astral influences with a celebration of the human capacity, not just to know but also to *govern* the universe. Man alone is not simply determined *by* nature, he is capable of determining the course of events and he does this in some way by exploiting his association with or knowledge of the astral deities.

The esoteric and highly mystical and intellectual Hermetic tradition thus reinforced the notions of engineers and artisans that knowledge and practice should go together and that man's greatest dignity is manifested in his capacity for creative activity. In fact, Hermeticism provided a deeply felt religious and spiritual sanctification of certain kinds of worldly activity, and made it the *duty* of adherents to seek knowledge of the world in order that they might manipulate the world. In its most extreme form, Christian Hermeticism suggested that it was within the realm of human power to reverse the effects of Adam's fall by joining God in "a holy and most solemn restoration of Nature herself."[7]

But the initial thrust of Hermeticism was not to generate support for the secular engineering tradition, with its focus on Archimedian applied mathematics and physics and their use in the construction of buildings, canals, fortifications, and mechanical apparatus. It was to encourage a remarkable flowering of astrology, and magical and "occult" arts.

One of the great feats attributed to Hermes by ancient tradition was the creation of "living" images of the Egyptian gods—that is, the creation of statues capable of speaking and moving. Whether or not some early member of the Hermetic cult actually created automata that could simulate speech and animal motions, we do not know (the technology would have been available in Alexandria during the second century). It is clear, however, from *Asclepius* and other ostensibly Hermetic works that Renaissance thinkers believed that Hermes had created terrestrial gods "by knowing the occult (hidden) properties of substances, by arranging them in accordance with the principles of sympathetic magic, and by drawing down into them the life of the celestial gods by invocations."[8]

It would be a serious error to claim that all of the natural magic and number mysticism of the Renaissance owed a direct debt to the Hermetic corpus. Instead, it seems that writers like Ficino and his student, Pico della Mirandola, fashioned out of the Hermetica a doctrine that encouraged the exploration of many independently existing traditions of magical and mystical studies—what we in the twentieth century call the "occult" sciences. The Jewish cabala, Pythagorean number magic, talismanic magic from Egypt and Mesopotamia, and assorted alchemical theories and astrological schemes were brought into the open and more widely studied than ever before in an attempt to discover the hidden secrets of the universe that could give man power to gain wealth, heal the sick, restore to nature her pristine glory, and generally exercise the power over the physical world that was his by divine bequest.

A great effort was made by Renaissance magi to disassociate themselves from the notions of black magic, which had long been linked to converse with evil demons. Instead, magic was conceived, as it appeared in innumerable book titles, as "natural" magic. That is, as a magic that depended for its efficacy on the discovery and exploitation of the hidden virtues, powers, and natural laws that God had placed in the universe.

In John B. della Porta's words, *Natural Magic* was nothing but the "survey of the whole course of nature."[9] Such natural magic even received the stamp of religious orthodoxy when, in 1493, Pope Alexander VI reversed Innocent VIII's condemnation and supported the Hermetic magus, Pico della Mirandola, against theological critics, declaring Pico to be a faithful son of the church, and absolving his works—which embodied all of the Hermetic attitudes we have mentioned—from any taint of heresy.[10] Official papal endorsement of astrology and magic did not last beyond the early Counter-Reformation decisions of the Council of Trent; but for a time during the late fifteenth century and the first two-thirds of the sixteenth century, astrological and magical concerns were rampant and, by definition, orthodox Christian concerns.[11]

One need not reflect very long to recognize that the aims and attitudes of natural magicians, whether Hermetics or not, must have been closely linked with those we chose to define science in chapter 1. Initially, magical and scientific activities diverged in at least two fundamental ways. The simpler way was the tendency of some magi to want to confine occult knowledge to some spiritual elite so it could not be wrongfully applied. Along with this notion often went an assumption that the efficacy of magical practices might depend upon the spiritual condition of the magus. Thus, magical theory often denied the *universality* of occult relations. Many sincere alchemists, for example, insisted that base metals could indeed be transmitted into gold, or that a universal medical remedy—the philosophers' stone—could be created according to some theory, in the face of repeated failures. They rationalized the failures by appealing to the supposed imperfection of the spiritual state of the alchemist.

Another way magic diverged from scientific attitudes was in focusing attention not on natural phenomena, but on words and numbers. Cabalistic magic, for example, depended on the assumption that the names of things had a special power and contained special secrets. The behavior of objects was thus supposed to be influenced by appeal to their names.

From its earliest Renaissance revival, however, much natural magic was scientific in many of its essential attitudes—it focused on phenomena, sought universal and testable knowledge, looked for the underlying causes of phenomena, and attempted to systematize the knowledge gained. Furthermore, much of this kind of magic, or occult science, had clear ties to older scientific traditions and activities. Alchemy, for example, depended on the Aristotelian theory of the transformation of elements; and alchemical practice called for empirical studies of the properties of substances. Celestial magic depended on positional astronomy, and called for empirical observation of the "influences" of celestial configurations on terrestrial events. And mathematical magic, or what Thommaso Campanella called "real artificial magic," was intimately bound up with the sciences of mechanics and mathematics.

Campanella's discussion of real artificial magic shows how closely magic and science were often linked by the late Renaissance.

> Real artificial magic produces real effects, as when Architas made a flying dove of wood, and recently at Nuremburg, according to Boterus, an eagle and a fly have been made in the same way. Daedalus made statues which moved through the action of weights or of mercury. However, I do not hold that to be true which William of Paris writes, namely that it is possible to make a head which speaks with a human voice as Albertus Magnus is said to have done. It seems to me possible to make a certain imitation of the voice by means of reeds conducting the air, as in the case of the bronze bull made by Phalaris which could roar. *This art cannot produce marvelous effects save by means of local motions and weights and pulleys or by using a vacuum, as in pneumatic and hydraulic apparatuses, or by applying forces to the materials.* But such forces and materials can never be such as to capture a human soul.[12]

This discussion is not a tongue-in-cheek denial of the magical character of certain productions. It is a sincere attempt to characterize a particular form of what was magic to him, but has been subsumed under the title of applied physics or engineering by us. In fact, the statement allows us to understand how magical preoccupations could easily motivate what we now see as scientific studies during the Renaissance.

Just as magic and mechanics were tied together in the attitudes of men like Campanella, astrological motives provided one of the powerful stimuli for astronomical advance during the Renaissance. Although Copernicus was probably little influenced by Hermetic and astrological concerns, there can be little doubt that one of Tycho Brahe's reasons for attempting—and achieving—remarkable improvements in observational astronomy, was astrological.[13] Similarly, there is evidence to believe that Johannes Kepler's early theoretical studies were deeply influenced by his number mysticism and his astrological beliefs.[14]

In no subject were Renaissance magical beliefs and aims, and scientific attitudes and activities, more closely linked than in the study of alchemy. And among alchemists no figure embodied those links more fully than Philippus Aureolus Theophrastus Bombastus von Hohenheim, known as Paracelsus (1493–1541), whose post-humously published works sparked almost endless debates among physicians and natural philosophers alike.

Above all things, Paracelsus was a hater of (and battler against) traditional medicine and "academic" science of all kinds. To some extent, his hatred was that of a lower-class country physician who saw, like Vesalius, that traditional medical training set up a group of doctors who knew much about books but little about actually curing sick people. In addition, his hatred was that of a reform-minded Christian who saw Aristotle and Avicenna as heathen philosophers whose works contravened scripture. The attitudes of the natural magicians appealed to him because they were often intensely religious, focused on activity in the world, and suggested that useful knowledge was to be sought directly from God's natural creation rather than from an aristocratic and anti-Christian tradition. So antiintellectual were his views that he largely ignored the mathematical emphasis that was part and parcel of most Renaissance magic. The thurst of Paracelsan attitudes can best be seen in the comments of Peter Severinus (1540–1602) who collected, edited, and published his master's works after his death. Severinus "told his readers that they must sell their possessions, burn their books, and begin to travel so that they might make and collect observations on plants, animals, and minerals. After their *Wanderjahren*, they must 'purchase coal, build furnaces, watch and operate with the fire without wearying. In this way and no other you will arrive at a knowledge of things and their properties.'"[15]

Lest this emphasis on observation and experiment seem thoroughly modern, it is worth considering one of the kinds of discoveries that Paracelsus sought in his observations. According to Paracelsus and many other magi, God placed in many natural objects signs by which the properties of those objects could be recognized. Thus, Paracelsus wrote,

> Behold the Satyrion root, is it not formed like the male privy parts? Accordingly magic discovered it and revealed that it can restore a man's virility and passion . . . and the *Syderica* bears the image and

form of a snake on each of its leaves, and thus, according to magic, it gives protection against any kind of poisoning.[16]

Of course God did not implant such obvious signs in all materials, so the search for the occult, hidden, or magical properties of herbs and minerals led directly to careful experiments to test the efficacy of specific doses of specific drugs on specific diseases—that is, to systematic experimentation. In this way, once again, magical attitudes and activities merged into scientific ones.

SYMPATHETIC MAGIC, RENAISSANCE ASTRONOMY, AND MEDICINE

A single set of ideas linked virtually all Renaissance magical and astrological thought with Aristotelian and Platonic cosmologies alike. This set of ideas is embodied in the phrases "sympathetic magic" and "microcosm-macrocosm correspondence." Until one understands the importance, meaning, and pervasiveness of these notions it is virtually impossible to realize the extent to which a change in scientific theorizing might have a dramatic impact on a wide range of Renaissance cultural attitudes and activities.

Eugenio Garin has nicely summarized the attitude that pervaded the occult, alchemical, and astrological traditions.

> The point of agreement . . . lay precisely in the idea that the universe was alive, that it was full of hidden correspondences, of occult sympathies, and that it was completely pervaded by spirits, all of them refracting signs pregnant with hidden meanings. In this universe every thing, every being, every force was like a voice not yet fully understood, like a word suspended in mid-air. It was a universe in which every word had innumerable echoes and resonances, in which the stars were sending messages and listening to us and to each other. This universe was an immensely multiple and varied form of speech, now soft, now loud, now in secret messages, now in disclosed sentences. In the middle there was man, a miraculous being, subject to change, a being capable of uttering all words, transforming all things, and drawing all characters, a being capable of responding to every call and calling every God.[17]

In order to get some handle on the mass of hidden relationships within the universe, two basic notions were taken from a variety of older astrological and alchemical theories and amplified to become the basis of a new synthesis. We have seen in previous discussions of Aristotelian cosmology and Islamic astrological theory that the universe was perceived in terms of concentric rings, hierarchically arranged. Within the three-level, or three-"order," scheme of Abu Ma'shar, for example, a central sphere of corruptible terrestrial elements was surrounded by a celestial spherical shell inhabited by the stars and planets; that sphere in turn was surrounded by the divine sphere of God's invisible Heavens. This whole system was unified by the fact that aspects of the interior spheres could be "influenced" by events or entities in the outer spheres. Within each of the two interior spheres or shells, moreover, there was another hierarchy. Fire and air stood above water and earth, for example. During the late Middle Ages, Christian scholars had also established hierarchies within the visible heavens. As Nicole Oresme wrote,

> As soon as God had created the heavens, He ordained and deputed angels who should move the heavens and who will move them as long as it shall please them.[18]

To one planetary shell were assigned the Thrones, to another the Archangels, and so on. Thus the hierarchy in the visible heavens mirrored the invisible, Divine heaven.

Within the context of Renaissance magic, the notion that the universe was comprised of a series of hierarchically arranged orders, each order hierarchically arranged to mirror the hierarchy of those "superior" to it in some sense, was transformed into the incredibly powerful and suggestive doctrine of microcosm-macrocosm correspondence. Though at first it was used primarily to illuminate and understand the relationships between astronomical bodies and chemical bodies—that is,

Gold	corresponds to the Sun
Silver	corresponds to the Moon
Mercury (Hg)	corresponds to Mercury
Iron	corresponds to Mars
Lead	corresponds to Saturn

and so on—schemes were soon developed to relate the parts of Man (microcosm) to the heavens (macrocosm);

Sun	corresponds to the Heart
Moon	corresponds to the Brain (hence Lunatic)
Mercury	corresponds to the Male Genitalia

and so on. Since each order corresponded to every other, one could combine the above sets of correspondences to produce a set of correspondences between chemicals and the parts of man:

Gold	corresponds to Heart
Silver	corresponds to Brain
Mercury	corresponds to Gonads

At this point, a second set of ideas associated with the notion of sympathetic magic becomes important. If one looks at the initial list of correspondences one can see that a kind of logic related to appearances makes them plausible. The bright yellowish appearance of gold is in some sense the counterpart of the bright, yellowish light of the sun; the moon appears "silvery" in color; and Mars and iron both have a reddish cast (iron is usually a bit rusted or oxidized). These similarities are "signs"—like those discussed by Paracelsus—that there are "sympathetic" magical influences connecting the corresponding bodies. To the medical alchemist, for example, the correspondence between gold and the heart was a sign to try gold compounds in the treatment of diseases of the heart, and the correspondence between Mercury and the genitalia was a sign to treat venereal diseases with mercury salves.

The crucial consequence of all schemes associating different orders of existence with one another was to inextricably link what otherwise might have been considered independent or autonomous sciences or arts. For macrocosm-microcosm correspondences directly linked such things as celestial, meteorological, mineralogical, botanical, and anatomical phenomena through sympathetic magic. Furthermore, the assumption of such correspondences indirectly tied various sciences together because a knowledge of the relations between entities and events in any single order presumably provided insight into the relationships between corresponding entities in any other.

As an example, the presumed links between astronomy and medicine, which originated in the works of Abu Ma'shar and Avicenna, ensured that most Renaissance physicians would be deeply interested in astronomy because of the astrological aids to medicine that it offered. Robert Westman has convincingly demonstrated the nature of this connection by showing that Renaissance physicians often specialized in mathematics and astronomy as arts students before going on to take higher university degrees in medicine;[19] and the teachers of mathematics and astronomy in the arts faculty very frequently went on to get medical degrees to teach and practice medicine.

When we recognize these linkages between astronomy and other fields, like that of medicine, it becomes easier to understand the broad impact of innovation in specialized fields. In particular, we can understand the very substantial resistance to Copernican reforms in astronomy. While there were very few reasons from within the science of astronomy itself to criticize Copernicus' work, to those who depended upon traditional astronomical theory to buttress their medical activities, any fundamental change in astronomy was a serious threat. Westman characterizes the situation as follows:

> While astrology combined the predictive function of the astronomer with the explanatory role of the natural philosopher in the person of the academic physician, the alliance was . . . essentially a conservative one. The astrologer who calculated horoscopes or sought improved tables of mean motions never tried to challenge the explanatory fundamentals of his discipline. He was, in short, parasitic on astronomy for his calculational tools and on natural philosophy for his concepts of force and cosmological order. Small wonder, then, that professors of mathematics and medicine who were largely responsible for producing the popular and numerous almanacs, prognostications and iatromathematical treatises [i.e., treatises linking mathematical astronomy with medicine] of the sixteenth century were neither receptive to cosmological innovation nor the habit which produced it.[20]

Given this situation, it is easy to understand why men who followed the usual pattern from astronomy into medicine did not fully accept Copernicanism between 1543 and 1600, and why the few men who did accept Copernicanism were deviants from the usual professional pattern.[21]

THE COPERNICAN REVOLUTION IN ASTRONOMY

Before we try to explore the broad impact of Copernican astronomy on early modern culture, a very short discussion of traditional astronomy and of Copernicus' reform of astronomical theory is necessary. Within the medieval and early Renaissance university, astronomy was taught as a branch of mathematics and was usually formally divorced from natural philosophy. Though there were exceptions, astronomers usually accepted the limits of their domain as they had been articulated in Simplicius' commentary on Aristotle's *Physics* in antiquity.

> It is the business of physical inquiry to consider the substance of the heaven and the stars, their force and quality, their coming into being and their destruction, nay, it is in a position even to prove the facts about their size, shape, and arrangement; astronomy, on the other hand, does not attempt to speak of anything of this kind, but proves the arrangement of the heavenly bodies by considerations based on the view that the heaven is a real cosmos, and further, it tells us of the shapes and sizes and distances of the earth, sun, and moon, and of eclipses and conjunctions of the stars, as well as of the quality and extent of their movements. Accordingly, as it is connected with the investigation of quantity, size, and quality of form or shape, it naturally stood in need, in this way, of arithmetic and geometry. *The things, then, of which alone astronomy claims to give an account it is able to establish by means of arithmetic and geometry.* Now in many cases the astronomer and the physicist will propose to prove the same point, e.g., that the sun is of great size or that the earth is spherical, but they will not proceed by the same road. The physicist will prove each fact by considerations of essence or substance, of force, or its being better that things should be as they are, or of coming into being and change; the astronomer will prove them by the properties of figures or magnitudes, or by the amount of movement and the time that is appropriate to it. Again, the physicist will in many cases reach the cause by looking to creative force; but the astronomer, when he proves facts from external conditions, is not qualified to judge of the cause, as when, for instance, he declares the earth or the stars to be spherical; sometimes he does not even desire to ascertain the cause, as when he discourses about an eclipse; at other times he invents by way of hypothesis, and states certain expedients by the assumption of which the phenomena will be saved. For example, why do the sun, the moon, and the planets appear to move irregularly? We may answer that, if we assume that their orbits are eccentric circles or that the stars describe an epicycle, their apparent irregularity will be saved; and it will be necessary to go further and examine in how many different ways it is possible for these phenomena

to be brought about, so that we may bring our theory concerning the planets into agreement with that explanation of the causes which follows an admissible method. Hence we actually find a certain person, Heraclides of Pontus, coming forward and saying that, even on the assumption that the earth moves in a certain way, while the sun is in a certain way at rest, the apparent irregularity with reference to the sun can be saved. *For it is no part of the business of an astronomer to know what is by nature suited to a position of rest, and what sort of bodies are apt to move, but he introduces hypotheses under which some bodies remain fixed, while others move, and then considers to which hypotheses the phenomena actually observed in the heaven will correspond.* But he must go to the physicist for his first principles, namely, that the movements of the stars are simple, uniform, and ordered, and by means of these principles he will then prove that the rhythmic motion of all alike is in circles, some being turned in parallel circles, others in oblique circles.[22]

This statement makes it very clear that, although astronomy was grounded in a few very general principles taken from natural philosophy, astronomical systems did not usually claim to address the "causes" of celestial motions. In this sense, late medieval astronomy was peculiarly well situated to be the beneficiary of nominalist attitudes toward knowledge, and did not have to be reformulated in any significant way in response to Ockhamist ideas. Its function was to produce hypotheses that could *save the appearances* without claiming anything about the reality underlying the appearances.

By the early sixteenth century almost all European astronomical calculating systems were variants on the scheme developed by Claudius Ptolemy and presented in the work that had been titled the *Almagest* (Great Book) by its Islamic translators and commentators. Ptolemaic astronomy was reintroduced into European scholarship by Georg Purbach (1423–1461) in his *New Theory of the Planets*. And Purbach's student, Regiomentamus (1436–1476), published an *Epitome of Ptolemy's Almagest*, which summarized its mathematical parts.

The Ptolemaic system of astronomy is geocentric (earth centered), and assumes that the solar year (the period from summer solstice to summer solstice, or from longest summer day to longest summer day) is constant. The system had initially been designed to account for such phenomena as the variable sizes of such bodies as the moon, Venus, Mars, and Mercury; the daily motion of the general framework of the so-called "fixed" stars; the differing periods of motion of the planets through the background of the fixed stars; and the variability of angular velocity of the planets within each cycle—a variability that was so great that the planets not only appeared to slow down and speed up, but even appeared to stop and back up (have retrograde motion) occasionally.

In order to account for such phenomena, Ptolemy employed a number of mathematical devices that call for some explanation. We begin by thinking that the simplest possible heavenly motion would be a constant angular motion in a circle centered on the earth. Suppose for a moment that S_1, S_2, S_3 and S_4 in figure 21 are fixed stars 90° from one another on the "sphere of the fixed stars." Suppose also that the rotational speed of the sphere of the fixed stars is constant. If Planet P moves with constant angular velocity on its circle centered on E it will take equal times to move from S_1 to S_2, from S_2 to S_3, from S_3 to S_4, and from S_4 to S_1 again, as viewed from E against the starry background. Furthermore, it will take the same period to move from S_1 around its circle and back to S_1 in successive revolutions—that is, its period of revolution will be constant. Lamentably, no heavenly body moves so simply; but the sun moves in such a way that a very

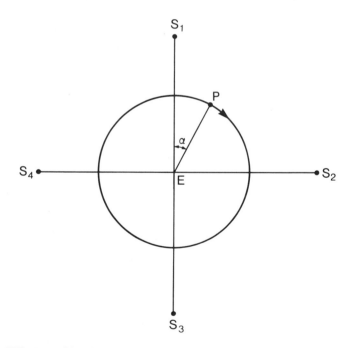

FIG. 21.   Simplest possible model of a planetary orbit ($\alpha = kt$).

minor modification of the simplest scheme can account for its apparent motion in a highly accurate manner.

In figure 22, suppose that S_1, S_2, S_3 and S_4 are successively the stars against which the sun appears at the spring equinox (equal day and night), summer solstice (longest day), autumnal equinox (equal day and night), and winter solstice (shortest day). The ancient astronomer, Hipparcus, knew that the periods of spring, summer, fall and winter—that is, from S_1 to S_2, S_2 to S_3, S_3 to S_4 and S_4 back to S_1—were as follows:

Season	Length	Paths in Fig. 22
Spring	94 ½ days	$S_1 S_2$
Summer	92 ½ days	$S_2 S_3$
Fall	88 ⅝ days	$S_3 S_4$
Winter	89 ⅝ days	$S_4 S_1$

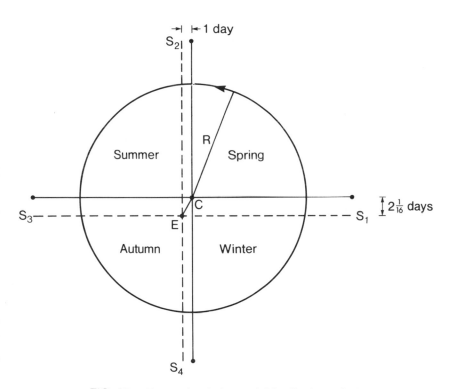

FIG. 22. Eccentric circle model for Sun's motion.

If we assume that the sun moves with a constant angular velocity about the point C, which is displaced from the center of the earth
as shown, then we can account for all of the observed phenomena.
The circular path assumed for the sun is called an *eccentric circle*
and the ratio of the distance EC to the radius R of the circle is called
the *eccentricity* of such a circle.

Mathematicians after Hipparcus showed that an eccentric
circle could be reproduced by another mathematical device, which
was to have great importance for Ptolemaic astronomy. Consider
figure 23. We begin with a circle centered on the earth (the deferent).
On this circle the center, C, of another circle (the epicycle) moves
with a constant angular velocity α. While C moves along the deferent,
the point P, representing the sun, revolves along the epicycle with an
angular velocity of $-\alpha$. In this way the point representing the sun
traces out a path identical to the eccentric circle of figure 22.

Now, if all we could do with deferents and epicycles was to
reproduce eccentric circles, their use would be totally uninteresting.
But consider figures 24 and 25, in which the angular motions of the
point P about the center of the epicycle C are respectively -2α and 8α.

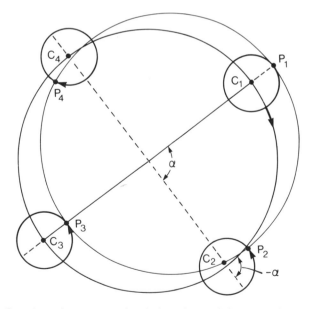

FIG. 23. Creation of an eccentric circle using a deferent and an epicycle.

In these cases the paths passed through by P are such that P comes closer to E near S_1 and S_3 than at S_2 and S_4 in figure 24 (i.e., the apparent size of P as viewed from E varies), and that P seems to slow down and reverse its motion (i.e., move in a retrograde motion) in figure 25.

Ptolemy used combinations of epicycles and deferents to account for several features of planetary motions, including the variable sizes and retrogradations. Unfortunately, no simple combination of a single epicycle with a single deferent was able to account for the motions of most heavenly bodies, so Ptolemy moved the center of the deferent away from the earth (i.e., made the deferent an eccentric circle) and/or added a second epicycle moving on the first to account for some motions—especially those of Mars and of the moon. Even this technique was unable to produce a system that fully accounted for all apparent motions; so Ptolemy adopted a final mathematical technique. He made the angular motion of the center C of the epicycle for some planets constant about another point, Eq, which we call the equant, and which is neither the Earth nor the center of the deferent (see figure 26). This technique allowed him to

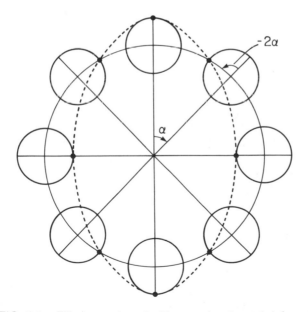

FIG. 24.   Ellipse produced with an epicycle and deferent.

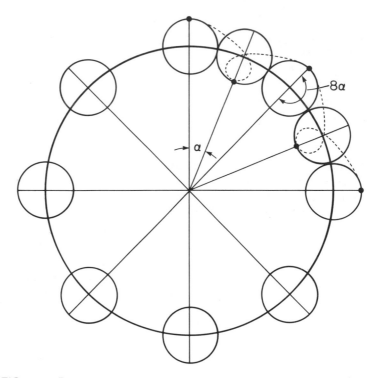

FIG. 25.   Retrograde motion produced with an epicycle and deferent.

produce the proper variations of apparent angular motion for all planets and at all points in their orbits.

Through the use of eccentric circles, deferents, epicycles, and equants, and with the proper choices of eccentricities, angular velocities, equant placements and initial positions, Ptolemy was able to provide a scheme for calculating the observed positions of heavenly bodies which indeed "saved the phenomena" during the period immediately after he lived. But by the period of the Renaissance, two classes of critical problems had arisen.

The first set of problems arose out of the fact that planetary, solar, and lunar positions as calculated according to Ptolemaic theory simply did not correspond to those actually observed; this produced problems for the religious calendar and for astrological prognostication. The calculated conjunctions of planets were frequently several days out of phase with observation, and Easter was out of phase by about ten days with respect to the spring equinox by

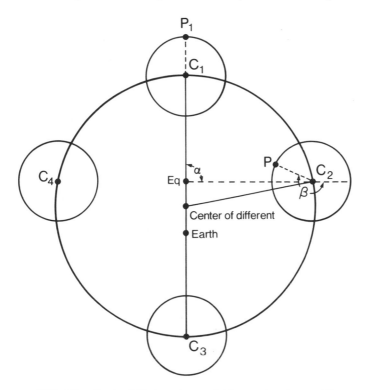

FIG. 26.   Typical Ptolemaic model for a planetary orbit
(α = angle C_1E_qC, β = 180° + angle E_qCP).

the early sixteenth century, causing the church to call a council on
calendar reform in 1512. At this meeting a young Polish astronomer,
Nicolas Copernicus, advised against calendar reform until astronomical theory could be revised.

A second set of problems for Renaissance astronomy grew out
of observations made in Islam. These observations suggested that
the relationship between solar and sidereal years was not constant.
A solar year is defined as the period between successive spring
equinoxes. A sidereal year is defined as the period between successive passages of the sun past a specified point on the sphere of the
fixed stars. Hipparcus and Ptolemy both realized that these two year
lengths were different, and that the point of the spring equinox moves
(precesses) through the fixed stars at the rate of about 1° per century.
Since the ratio between the two year lengths was presumed to be

constant, however, there was no problem in moving from positions measured relative to the sun (synodic) and those measured relative to the stars (sidereal). The sidereal period of any motion was simply the synodic period times the ratio of sidereal to solar years.

In fact, the sidereal and solar years *are* nearly constantly related, but a series of erroneously recorded Islamic and ancient observations indicated that they were not; and neither Copernicus nor any of his near contemporaries doubted the validity of those observations.[23] Consequently, either one had to assume that the solar year varied over time or that the sidereal year varied (i.e., that the motion of the sphere of the fixed stars was inconstant). Ptolemy made the first assumption, and later Ptolemaic astronomers tried to account for the variation (trepidation) of the motion of the fixed stars. Figure 27 indicates a mechanism proposed by Purbach, for example.

Copernicus seems to have been particularly bothered by the Ptolemaic assumption of constant solar years and hence variable motion of the fixed stars. Certainly, such an assumption violated all earlier arguments about the peculiar perfection and constancy of stellar motion that came from natural philosophy—whether Platonic

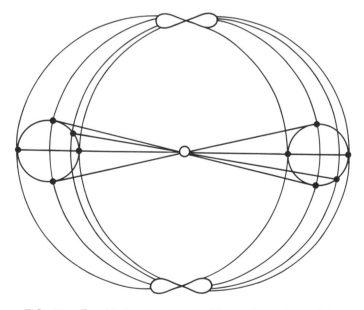

FIG. 27.   Trepidation is accounted for as the sphere of the
fixed stars moves on "idle wheels."

or Aristotelian. But even more critically, Copernicus felt that since virtually all motions of planets—including the motion of the sun itself—were observed against the background of the fixed stars, it was self-contradictory to suppose that the "fixed" reference system moved with an inconstant velocity.[24] For whatever reasons, in all of his attempts at astronomical reform Copernicus began from the assumption of a constant sidereal year. "Equal Motion should be Measured Not by the Equinoxes but by the Fixed Stars,"[25] he wrote.

Once set on the task of revising astronomical theory, Copernicus took the natural humanistic tack of reviewing older theories.

> I therefore took pains to read again the works of all the philosophers on whom I could lay hand to seek out whether any of them had ever supposed that the motions of the spheres were other than those demanded by the mathematical schools. I found first in Cicero that Hicetas had realized that the earth moved. Afterwards I found in Plutarch that certain others had held the like opinion. I think fit here to add Plutarch's own words, to make them accessible to all:—
>> "The rest hold the earth to be stationary, but Philolaus the Pythagorean says that she moves around the (central) fire on an oblique circle like the Sun and Moon. Heraclides of Pontus and Ecphantus the Pythagorean also make the earth to move, not indeed through space but by rotating round her own centre as a wheel on an axle from West to East."
> Taking advantage of this, I, too, began to think of the mobility of the Earth; and though the opinion seemed absurd, yet knowing now that others before me had been granted freedom to imagine such circles as they chose to explain the phenomena of the stars, I considered that I also might easily be allowed to try whether, by assuming some motion of the Earth, sounder explanations than theirs for the revolution of the Celestial spheres might so be discovered.[26]

The preliminary version of Copernicus' new theory appeared in *The Commentariolus* which was composed sometime before the mid-1520's, and while the masterful *De Revolutionibus* somewhat modifies and greatly extends the approach taken in *The Commentariolus*, the earlier text contains the most fundamental nontechnical aspects of Copernicus' works in a form that is much more readable for nonmathematicians than the great masterpiece. In particular, it presents the essence of Copernicanism in the form of seven basic assumptions:

1. There is no one center of all the celestial circles or spheres.
2. The center of the earth is not the center of the universe, but only of gravity and of the lunar sphere.

3. All the spheres revolve about the sun as their mid-point, and therefore the sun is the center of the universe.
4. The ratio of the earth's distance from the sun to the height of the
   firmament is so much smaller than the ratio of the earth's radius to
   its distance from the sun that the distance from the earth to the sun
   is imperceptible in comparison with the height of the firmament.
5. Whatever motion appears in the firmament arises not from any
   motion of the firmament, but from the earth's motion. The earth
   together with its circumjacent elements performs a complete rotation on its fixed poles in a daily motion, while the firmament and
   highest heaven abide unchanged.
6. What appear to us as motions of the sun arise not from its motion
   but from the motion of the earth and our sphere, with which we
   revolve about the sun like any other planet. The earth has, then, more
   than one motion.
7. The apparent retrograde and direct motion of the planets arises not
   from their motion but from the earth's. The motion of the earth alone,
   therefore, suffices to explain so many apparent inequalities in the
   heavens.[27]

The resulting system (for motions along the ecliptic) can be
presented as in figure 28.

If we consider the Copernican and Ptolemaic systems solely as
mathematical calculating systems, they are equivalent systems for
predicting observed positions of the planets, the moon and the sun.
Because Copernicus had the advantage of establishing new initial
positions and refined period relations, his system produced more
accurate tables of position than Ptolemy's; but that was not the result
of a changed *system* of calculation.

The Copernican scheme did, however, offer some advantages
over that of Ptolemy. Though it involved about the same number of
deferents and epicycles, it managed to avoid using equants. Furthermore it (1) allowed for a more natural explanation of retrograde
motions based on the differential velocities of the earth and other
planets; (2) provided a natural explanation of the fact that the
observed periods of the sun, Venus, and Mercury were the same;
(3) provided a natural progression from the fastest to the slowest
motions as the distance from the sun increased; and (4) provided a
means for determining the relative sizes of planetary orbits.

From a strictly mathematical point of view, only two major
inconveniences accompanied these gains. First, a mathematical
theory that posits only a single center of all motions is aesthetically
preferable to one that posits several centers—so assumption one of
The Commentariolis is a net loss from the mathematical point of

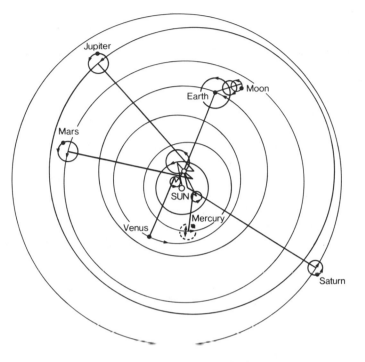

FIG. 28.   Copernican cosmological system.

view. Second, since all positions are observed from the earth it is necessary in Copernican astronomy to do some time-consuming trigonometric calculations to move to positions as viewed from the sun and back—but this is a purely computational inconvenience.

Had astronomers been able to evaluate the Copernican system strictly as a calculating scheme with no broader implications for natural philosophy and other topics tied to cosmology, they probably would have rapidly accepted it. But it seems clear that Copernicus neither intended his work to be devoid of philosophical implications,[28] nor that many readers were able to read it without seeing that if accepted, it would directly overturn virtually all traditional natural philosophy and cosmology and indirectly disrupt the rationales for such overarching schemes as those associated with macrocosm-microcosm correspondences.

Andreaus Osiander, who took on the task of publication of *De Revolutionibus* for an aging and ill Copernicus, foresaw the likely problems and tried to forestall them by adding an anonymous "letter to the reader" denying that any philosophical implications should be imputed to the work, which was to be viewed solely as a calculating scheme based on assumptions that were not supposed to reflect reality. "I have no doubt," he wrote, "that some learned men have taken serious offence because the book declares that the earth moves and that the sun is at rest in the center of the universe; *these men undoubtedly believe that the liberal arts, established long ago on a correct basis, should not be thrown into confusion.*"[29]

Such men should have no fear, Osiander continued, "For it is quite clear that the causes of apparent unequal motions are completely and simply unknown to this art. And if any causes are devised by the imagination, as indeed very many are, they are *not* put forward to convince any one that they are true, but merely to provide a correct basis for calculation."[30]

Regardless of Osiander's claims, most readers could not read Copernicus without seeing that his work denied, among other long-standing assumptions, the fundamental presumption of a hierarchical spatial ordering of the cosmos from the highest Divine sphere to the lowest terrestrial sphere. Consequently, the largest potential audience for Copernicus' work, the class of medical-astrologers, found Copernicanism incompatible with their most fundamental beliefs; and they naturally rejected it.

POLITICAL AND THEOLOGICAL IMPLICATIONS OF COPERNICAN THEORY

If the traditional cosmological assumptions challenged by the Copernican theory had been limited to the use of macrocosm-microcosm correspondences in a handful of magical arts—even such important ones for Renaissance life as astrological medicine and alchemy—they would not deserve our close attention here. But that was not the case.

In the first place, in 1563 the Council of Trent reestablished St. Thomas' focus on natural theology and the necessity of making scientific and theological opinions compatible. This not only ensured that churchmen—especially the Jesuit order, which took the most active role in Christian education during the early modern era, and the Dominicans, who were entrusted with defending church dogmas—would pay attention to cosmological writings, but that they would actively consider the relation of such writings to Christian dogma.

Even more importantly, the tendency to view microcosmic-macrocosmic correspondences as the foundation of the understanding of all orders of existence—especially of the political and social orders of human existence—was of overwhelming importance during the Renaissance. And this tendency inextricably linked cosmological and political attitudes in such a way that a series of simple technical changes in astronomy could seem to threaten both the ability to understand politics and society, and the very stability of society itself.

Images of correspondence between heavenly hierarchies and social ones, and between the microcosm, man, and the macrocosm, state, provided the central arguments of conservative political ideology during the late sixteenth and very early seventeenth centuries. "This is Nature's nest of boxes," wrote John Donne in one of his sermons, "The Heavens contain the Earth, the Earth cities, Cities, Men. And all these are concentric."[31] From this concentricity one recognized how the plan of each successive internal order mirrored the plans of those superior to it.

This notion provided a rationale for aristocratic social and political hierarchies for the authors of late Renaissance court literature, and for those who defended a hierarchical structure of church government against those who sought a more democratic arrangement. But to illustrate the pervasiveness and power of the linkage between celestial and social orders there is no better place to start than with the famous speech of Ulysses from Shakespeare's *Troilus and Cressida*, written shortly before 1603.

In this speech, Ulysses explains to his Greek countrymen why they have failed to capture Troy in spite of their supposed superiority in numbers and military training. Greek military discipline has broken down and confusion has arisen because the great Greek general, Achilles, has not behaved in a manner proper to his station.

The specialty of rule hath been neglected.
And look how many Grecian tents do stand
Hollow upon this plain, so many hollow factions.
When that the general is not like the hive
To whom the foragers shall all repair,
What honey is expected? Degree being
       obscured . . .
The heavens themselves, the planets and this
       center,
Observe degree, priority, and place,

Insistiture, course, proportion, season, form,
Office, and custom, in all line of order.
And therefore is the glorious planet Sol
In noble eminence enthroned and sphered amidst
            the others, whose medicinable eye
Corrects the ill aspects of planets evil,
And posts like the commandment of a king,
Sans check to good and bad. But when the planets,
In evil mixture to disorder wander,
What plagues and what portents, what mutiny
What raging of the sea, shaking of the earth,
Commotion in the winds, frights, changes,
            horrors,
Divert and crack, rend and deracinate,
The Unity and married calm of states
Quite from their fixture! Oh, when degree is
            shaked,
Which is the ladder to all high designs,
The enterprize is sick! How could communities,
Degrees in schools and brotherhoods in cities,
Peaceful commerce from dividable shores,
The primogenitive and due of birth,
Prerogative of age, crowns, sceptres, laurels,
But by degree, stand in authentic place?
Take but degree away, untune that string,
And hark, what discord follows! Each thing meets
In mere oppugnancy. The bounded waters
Should lift their bosoms higher than the shores,
            and make a sop of all this solid globe.
Strength should be lord of imbecility,
And the rude son should strike his father dead . . .
    . . .
Thus chaos, when degree is suffocate,
Follows the choking
And this neglection of degree it is
That by a pace goes backward, with a purpose
It hath to climb. The general's disdained
By him one step below, he by the next,
That next by him beneath. So every step,
Exampled by the first place that is sick
Of his superior, grows to an envious fever

Of pale and bloodless emulation.
And 'tis this fever that keeps Troy on foot,
Not her own sinews. To end a tale of length,
Troy in our weakness stands, not in her
        strength.[32]

In this manner did Shakespeare, whose mind was a sponge for widely shared late Renaissance ideas and whose artistic ability made him a powerful spokesman for them, express the conservative belief in the absolute necessity of a hierarchically ordered social structure, and the authorization or rationale that belief drew from the cosmological order embodied in traditional astronomy and natural philosophy. To challenge those hierarchical structures at any level could not fail to have widespread ramifications for men who held such beliefs.

It is important to realize that these microcosm-macrocosm analogies were not merely meaningless rhetorical conventions. They were conventions—what Samuel Johnson later denigrated as "conceits"—but they both mirrored and produced real convictions. This fact is made peculiarly obvious by the physician and poet, Thomas Browne, who at first encountered them as simple poetic devices, but who became obsessed with their significance as time went on. "That we are the breath and similitude of God, it is indisputable, and upon record of Holy Scripture;" he wrote in *Religio Medici* (1643), "but to call ourselves a Microcosm, or little World, I thought only a pleasant trope of Rhetorick, till my neer judgement and second thoughts told me there was a real truth therein."[33] In his later work, Browne endlessly sought to explore the relations between man, the microcosm, and the world about him.

Turning to didactic and overtly political writings, Thomas Elyot illuminates a variety of issues connected with the correspondence between celestial, political, and human orders in *A Book called the Governor* (1531), which was not only the first major English didactic work on political theory, but the book that introduced such words as *beneficence, clemency, education, equability, frugality, imprudence, liberty of speech, loyalty, magistrate, mediocrity, sincerity, society,* and above all, *democracy,* into the English language.[34] *The Governor* was in many ways patterned on Plato's Republic. It sought first to explain, "the form of a just public weal," and then, like the *Republic,* it sought to treat "the education of them that hereafter may be deemed worthy to be governors of the public weal," focusing thus on the education of the governing class.[35] In order to carry the burden of the argument in favor of an aristocratic society headed by a monarch and an intellectual elite, however, Elyot departs from Plato to appeal to the theory of correspondences.

> Hath not God set degrees and estates in all his glorious works? First in his heavenly ministers, whom he hath constituted in diverse degrees called hierarchies. Behold the four elements, whereof the body of man is compact, how they be set in their places called spheres, higher or lower according to the sovereignty of their natures. Behold also the order that God hath put generally in all his creatures, beginning at the inferior, or base, and ascending upward. He hath made not only herbs to garnish the earth but also trees of a more eminent stature than herbs. . . . Every kind of trees herbs birds beasts and fishes have a peculiar disposition apportioned to them unto them by God, their creator; so that in everything is order, and without order be nothing stable or permanent. And it may not be called order except it do contain in it degrees, both high and base, according to the merit or estimation of the thing that is ordered . . .
>   And like as Angels which he most fervent in contemplation be highest exalted in glory (after the opinion of holy doctors), and also the fire which is the most pure of elements is deputed to the highest sphere or place; so in this world they which excel others in this influence of understanding and do employ it to the detaining of others within the bounds of reason and show them how to provide for their necessary living, such ought to be set in a more high place than the residue where they may see and also be seen.[36]

Sir Walter Raleigh, less concerned with intellectual "nobility," but even more concerned to rationalize the existing power relations in Elizabethan society, identifies wealth and social rank as the crucial determinants of the human hierarchy.

"The common people are evil judges of honest things, and whose wisdom (saith Ecclesiastes) is to be despised,"[37] he writes, and after acknowledging the importance of piety and wisdom he goes on,

> Shall we therefore value honor and riches at nothing? And neglect them, as unnecessary and vain? Certainly no. For that infinite wisdom of God, which hath distinguished his angels by degrees; which hath given greater and less light and beauty to heavenly bodies; which hath made differences between beasts and birds; . . . and among stones, given the fairest tincture to the ruby and the quickest light to the diamond; hath also ordained kings, dukes, or leaders of the people, magistrates, judges, and other degrees among men.[38]

An important point needs to be made here. I do not want to argue that the aristocratic character of late Renaissance society or the focus on strong monarchs came about because men intentionally sought to emulate celestial hierarchies in their social organizations. We can and must account for the basic features of Renaissance society and the growing power of monarchical states primarily in terms of social and economic conditions. It is undoubtedly the case, however, that political ideologies—that is, theories that provide a rationale for, and in some sense legitimize, the existing social order—have some significant conservative social impact. They defuse and diffuse social unrest and serve to generate support from those whose real economic and social interests are *not* served adequately by the existing institutions. By the same token, Utopian social theories—theories that rationalize social change—tend to channel and focus social unrest and to erode the confidence and authority of those in power. Conservative political thought may thus retard the rate of social change; and radical or Utopian political theory, though it may have little influence on the general direction of change, can often have substantial impact on the rate of change and on the specific institutional forms in which that change manifests itself.

It would be impressive at this point to be able to claim and demonstrate that the Copernican reformation of astronomy, with its undermining of traditional celestial hierarchies, provided an immediate and direct impetus to secular political theories that encouraged and accelerated more republican and democratic tendencies. To my knowledge, that did not happen. But an attack on celestial hierarchies with direct allusions to Copernican support *did* play a significant role in late sixteenth- and early seventeenth-century attacks on traditional church (or ecclesiastical) governance, which in turn provided an impetus to later secular theorizing. Moreover, conservative defenses of ecclesiastical hierarchies were often formulated in terms of an appeal to the correspondences between the natural order and the order of the church.

In order to understand some of the issues at stake among Reformation challenges to the traditional church hierarchy, I will focus on theological and intellectual issues related to notions of celestial and corresponding ecclesiastical hierarchies. I do this not because I think social, economic, and political conditions are irrelevant, but because I want to consider the significant—though not totally determinative—role of scientific developments in shaping men's *understanding* of institutional realities.

Renaissance theologians' ideas regarding the correspondence between celestial hierarchies and church government were deeply influenced by three writings attributed to St. Dionysius the Areopagite: the *Celestial Hierarchies*, the *Ecclesiastical Hierarchies* and the *Mystical Theology*. Ficino, who was responsible for a renewed emphasis on this set of works, which had been known since late antiquity, believed that St. Dionysius was a contemporary of St. Paul and a source for Pauline doctrine. As in the case of Hermes, Ficino was wrong; the writings are probably fifth-century Neoplatonic writings. But once more, for our purposes, it is important to recognize that they seemed in the Renaissance to gain authority by their supposed association with St. Paul. The basic thrust of these works was twofold. At one level—which we virtually ignore—they encouraged an extreme mysticism within Christianity and heightened the importance of allegorical interpretations of scripture. At a second level—which we focus on—they argued for the necessity of a hierarchical structure of institutions to mediate between ordinary men and an extremely distant, hidden God.

Corresponding to the nine spheres associated with the material cosmos, Dionysius claimed scriptural authority to posit the existence of nine orders of immaterial spiritual beings, whose functions are indicated by Ficino as follows:

> Seraphim speculate on the order and providence of God.
> Cherubim speculate on the essence and form of God.
> Thrones also speculate, though some descend to works.
> Dominions, like architects, design what the rest execute.
> Virtues execute, and move the heavens, and concur for the working of miracles as God's instruments.
> Powers watch that the order of divine governance is not interrupted and some of them descend to human things.
> Principalities care for public affairs, nations, princes, magistrates.
> Archangels direct the divine cult and look after sacred things.
> Angels look after smaller affairs and take charge of individuals as their guardian angels.[39]

Moreover, just as it was assumed by cosmologists that the motion of inner spheres was communicated to them from those immediately above rather than directly from the "Primum Mobile," Dionysius assumed that the contact of lower orders of spiritual beings with God was transmitted through the higher orders. Thus, he wrote,

> The highest Order, as we have said, being in the foremost place near the Hidden One, must be regarded as hierarchically ordering in a hidden manner the second Order; and the second Order of Dominions, Virtues, and Powers, leads the Principalities, Archangels and Angels more manifestly, indeed than the first Hierarchy, but in a more hidden manner than the Order below it; and the revealing Order of the Principalities, Archangels, and Angels presides one through the other over the human hierarchies so that their evolution and turning to God and their communion and Union with him may be in order . . .[40]

Continuing this line of argument in *The Ecclesiastical Hierarchy*, Dionysius discusses a nine-level hierarchy within the church, moving downward from the three major sacraments—baptism, unction and the Holy Eucharist—through three levels of clergy—the Bishops, Priests, and Deacons—to the three levels of laity—those in monastic orders, those who have partaken of the sacraments, and those preparing for baptism or who have lapsed and must be readmitted to the second rank. Again, as in the case of the cosmic and spiritual hierarchies, members of lower orders can approach the divine only through the mediation of the orders above.

> Thus the order of bishops is that which enjoys the fulness of the power of administering the sacraments . . .
> The order of priests, whose function is illumination, conducts the initiate to initiation into the sacraments in subordination to the divine order of bishops . . .
> Finally, the order of deacons [ministers] is charged with the duty of purifying and sifting those who do not bear them the resemblance of the divine, before they can approach the rites of the priests . . .[41]

Though the church of the Renaissance had expanded or subdivided the orders within the church, it remained the case that the church's fundamental claim to being the necessary mediator between men and God was often rationalized through this set of ancient cosmological, spiritual, and ecclesiastical correspondences.

Since it was the fundamental aim of Protestant reformers to return to a presumed "primitive" Christianity in which men need no intervening institutions to mediate between themselves and God, it was absolutely necessary to undermine the Dionysian presupposi-

tion. Thus, John Calvin specifically directed his voice against
Dionysius.

> None can deny that Dionysius (whoever he might have been) has many
> shrewd and subtile disquisitions in his *Celestial Hierarchy*; but only
> looking at them more closely, everyone must see that they are idle
> talk . . .[42]

Calvin went on to deny that God worked primarily through
angelic beings, and he absolutely denied the propriety of thinking of
church governance in terms of some special hierarchy.

> To the government thus constituted, some gave the name Hierarchy—a
> name in my opinion improper, certainly not one used by Scripture. For
> the Holy Spirit designed to provide that no one should dream of pri-
> macy or domination in regard to the government of the church.[43]

That the new Copernican system of the world undermined the
older notion of hierarchical structure and could be used to authorize
a more "democratic" attitude in which all humans were in direct
communion with God was probably recognized at least by some
reform-oriented scientists. The central sun rather than the distant
prime mover becomes for them the abode of God, and all the planets
become equals. Juachim Rheticus, Copernicus' Protestant student,
hinted at this change in his *Narratio Prima* of 1539, in which he
points out that, "In the hypothesis of my teacher, the sphere of each
planet advances uniformly with the motion assigned to it by nature,
and completes its period *without being forced into any inequality
by the power of a higher sphere.*"[44] But the point was made in
explicitly antihierarchical and political terms by William Gilbert,
who adopted some, but not all, elements of Copernicanism. "The sun
is not swept round by Mars' sphere (if sphere we have) and its
motion," wrote Gilbert, "nor Mars by Jupiter's sphere, not Jupiter by
Saturn's. *The higher do not tyrannize over the lower.*" And he went on
to assert that, "The earth's motion is performed with as little effort as
the motion of the other heavenly bodies . . . *neither is it inferior in
dignity.*"[45]

The Protestant Kepler's acceptance of Copernicanism was likewise based not only on its mathematical and technical character, but
on its appeal to the unique, central, and unmediated power of the
sun.

The sun *alone* appears by virtue of his dignity and power, suited for this motive duty [of moving the planets] and worthy to become the home of God himself. . . . For if the Germans elect him as Caesar who has the most power in the whole empire, who would hesitate to confer the votes of the celestial motions on him who already has been administering all other movements and changes by its benefit of the light *which is entirely his possession?*[46]

It should hardly be surprising, then, that some Protestant humanists linked the name of Copernicus with those of Wycliff, Luther, Zwingli and Calvin. The religious reformers were not inventors but restorers who "restored it [the Church] to its purity, according to God's word," just as "Nicolaus Copernicus . . . restored to us the place of the starres according to the truth."[47]

It was not the case that all Protestant theologians went out of their way to learn about and adopt Copernican astronomical theory. There were obvious incompatibilities between literally interpreted Biblical statements and Copernicanism; and Calvin himself pointed these up in his *Commentary on Genesis*. Citing the opening verse of the ninety-third psalm—"the earth also is stablished, that it cannot be moved"—Calvin asked, "Who will venture to place the authority of Copernicus above that of the Holy Spirit?"[48] Nor was it the case that most—or even many—astronomers showed any indication that some presumed ecclesiastical or political implications of Copernican doctrine played a significant role in their theory choice. None the less, in the small but significant number of cases in which Copernicanism, or heliocentrism in general, was linked positively to ecclesiastical doctrine, the linkage was usually made by reformers who opposed the hierarchical structure of the church. Richard Overton, for example, one of the most active leaders of the Leveler movement during the English civil wars, insisted that the sun, which, "according to the famous Copernicus . . . is the highest station in the whole Creation," must be the abode of God, and that there were no adequate grounds for assuming the existence of a "Coelum Emperium which the Astronomers have invented for his residence" and from which he was supposed to govern the universe through a hierarchy of intermediaries.[49]

As Copernicanism became more and more identified with heretical religious viewpoints at the beginning of the seventeenth century, a strong current of anti-Copernican sentiment developed within the Catholic church. The church had based its calendar reform of 1586 on astronomical tables derived from Copernican theory; and the scientifically oriented Jesuits had quietly begun to study and teach at least the technical details of Copernican astronomy. But the more traditionally oriented Dominicans did not approve. Dominican sensitivity to the dangers of Copernicanism was particularly high in the first decade of the seventeenth century because the order itself had given birth to two popular and heretical figures, Giordano Bruno and Thomasso Campenella, both of whom tied their ideas, which were heretical independently of their astronomical concerns, to Heliocentric astronomical thought. Bruno, who went beyond mere Copernicanism to claim that the universe was infinitely large and that there were an infinity of universes, was incarcerated for eight years by the Inquisition (a Dominican institution within the Papal Court) and burned at the stake in Rome in February of 1600. Campenella led a Calabrian (south Italian) revolt against Spanish domination in the name of religious reform and the establishment of a "City of the Sun," modeled on a Heliocentric cosmological scheme. He was thrown into a Neopolitan prison at the end of 1599.

THE STRANGE CASE OF GALILEO AND THE INQUISITION

In light of these links between heresy and heliocentrism it should hardly be surprising that in 1613, when the mathematician and astronomer, Galileo Galilei, openly advocated adoption of the Copernican system as *true*, and not merely as a handy calculating system, he ran into a strong Dominican resistance movement. Galileo initially had support from the Jesuit astronomers and liberal-minded humanists within the church. In particular, he was befriended by Cardinal Mafeo Barbarini, who was to become Pope Urban VIII in 1623, and whose first acts included freeing Campanella from prison. But as early as 1614 a Dominican counter attack was underway. The Dominican friar, Thomasso Caccini, preached a forceful sermon against the Copernicans and Galileans under the catchy title, "Ye Men of Galilee, why stand ye gazing up into heaven." In this sermon he gathered together passages on Joshua and other literalist arguments often—and usually wrongly—attributed to fundamentalist Protestants.

Galileo responded to these attacks in a private letter to Benedetto Castelli. In this letter he tactlessly, but shrewdly, warned religious conservatives that "physical effects placed before our eyes by sensible experience should not *in any circumstances* be called in doubt by passages in Scripture that verbally have a different semblance, since not everything in Scripture is linked to such severe obligations as is every physical effect."[50]

A copy of this letter found its way into the hands of Niccolo Lorini, the Dominican professor of ecclesiastical history of Padua; and Lorini forwarded it to the Inquisition along with his own request that the leaders of the Copernican movement be forced to terminate their dangerous heretical teachings.

Though Galileo was cleared at this point of any wrongdoing beyond that of showing poor judgment, the investigation led to formal condemnation of the basic Copernican propositions that the earth moved and the sun stood still, and to the placing of Copernicus' *De Revolutionibus* and several other Copernican books on the Index, or list of prohibited books.

On February 26, 1616, Galileo appeared before Cardinal Belarmine who informed him verbally and in writing that he was no longer to hold the Copernican theory as truth. The written warning given to Galileo did *not* forbid considering the Copernican view as a mere mathematical hypothesis. For eight years Galileo maintained a discrete public silence regarding his Copernicanism, but when Barbarini became Pope, Galileo had a series of audiences with him and began work on what was to become his most famous work and the greatest defense of Copernican astronomical theory, the *Dialogue concerning the two Chief World Systems*. In this book, while Galileo technically maintained that he held to Copernicanism only as an hypothesis, he mercilessly twitted the upholders of traditional Aristotelian cosmology, forcefully advocated Bruno's heretical claims regarding the plurality of worlds, and made an ironic joke out of Urban VIII's demands that he show at least formal acceptance of church authority. Somehow, the book received the official approval of the bishop of Florence and was printed early in 1632; but as soon as it was printed, the Dominicans, now assisted by the Pope, who felt he had been double-crossed by his former friend, demanded that publication be stopped and that Galileo be tried as a heretic.

The details of Galileo's trial, which began on 12 April 1632, are fascinating. It seems clear, for example, that the prosecution's case depended on forged documents, which were recognized as false even by some of the Inquisitors, three of whom refused to sign the sentence.[51] Whatever justices or injustices were committed during the trial, on June 22, 1633, Galileo formally renounced the Copernican belief.

> I have been pronounced by the Holy office to be vehemently suspected of heresy—that is to say, of having held and believed that the sun is the center of the world and immoveable, and that the earth is not the center, and moves. Therefore, desiring to remove from the minds of your eminences, and of all faithful Christians, this strong suspicion reasonably conceived against me, with sincere heart and unfeigned faith I abjure, curse, and detest the aforesaid errors and heresies, and generally every other error and sect whatsoever contrary to the said Holy Church . . .[52]

Two days later, Galileo was given his sentence—life imprisonment in Rome—a sentence which was soon commuted to life under house arrest in Tuscany.

Within a few days, Galileo's fate was communicated to his friends throughout Europe, putting a very real damper on Italian and Catholic scientific activity in general. Giorgio de Santillana no doubt overstates the case when he writes, "We can date back to that day, indeed, the time when Florentine civilization, which had carried the world since the thirteenth century, practically vanished from history."[53] But there can be no doubt that the church's handling of Galileo served as an important symbol and a source of serious reflection. René Descartes, hearing of Galileo's trial, decided to postpone publication of his radical scientific work, *Le Monde*. The Protestant apologist for speculative freedom, John Milton, wrote in 1644 of his visit to Italy,

> I have sat among their learned men . . . and been counted happy to be born in such a place of philosophic freedom as they supposed England was, while themselves did nothing but bemoan the servile condition into which learning amongst them was brought; that this was it which had damped the glory of Italian wits; that nothing had been there written now these many years but flattery and fustian. There it was that I found and visited the famous Galileo, grown old, a prisoner of the Inquisition, for thinking in astronomy otherwise than the Franciscan and Dominican licensers thought.[54]

Intellectual hegemony no doubt moved away from Italy because northern Europe became the center of economic power as the old Florentine cloth trade moved to Flanders and England. But the treatment of Galileo accelerated the Italian decline and provided an object lesson for Protestant apologists for freedom of speech and thought to use in trying to establish those doctrines as guiding principles wherever they might live. Curiously, it seems to have had a minimal impact within the Catholic church, whose Jesuit scholars continued to read and teach Copernican mathematical astronomy and to play a major role in early seventeenth-century scientific developments.[55]

From Renaissance to Modern Scientism in the Works of Johann Andreae and Francis Bacon

In the last two chapters I tried to show that scientific knowledge and attitudes became increasingly associated with practical traditions of artistic production and technological development during the Renaissance, and that this association had the reciprocal effects of scientizing technology and turning scientific knowledge increasingly toward direct practical operations. Simultaneously, it began to give artisans intellectual interests and pretensions, and turned some intellectuals toward greater interest in the manual arts and practical productive activities. In general, it undermined old notions that asserted the greater value and dignity of leisured gentlemen over that of productive artisans. While all of this was going on, a second tradition, which linked theology, magic, alchemy and astrology, sanctified and reemphasized the human capacity to manipulate nature and to serve human needs. This tradition intensified linkages between cosmological, political, and religious beliefs, and provided new religious and magical motives for the study of natural phenomena. Ties between cosmology, secular politics, and theological issues became particularly obvious when the political and theological emphasis on macrocosm-microcosm analogies and the ramifications of the Copernican revolution in astronomy were considered.

By the beginning of the seventeenth century these traditions began to receive a new focus and integration in the activities and writings of a few men. Thomasso Campanella, for example, was clearly tied to the magical tradition, the mechanical artisan tradition, and the Copernican tradition; and he linked all of these concerns with radical political and religious attitudes. Similarly, Francis Bacon was deeply involved in popularizing and extending the scientistic emphasis involved in Renaissance movements. But neither Campanella nor Bacon present ideal illustrations of the broad dispersion of the Renaissance attitudes. Campanella was too much of a mystic, magus, and political radical to seem characteristic; and Bacon too thoroughly rejected magic, the centrality of mathematics, and Copernicanism.

There probably was no *typical* response to the combination of Renaissance traditions. But there was one man who was respected and admired by adherents of virtually all the positions discussed so far (except for the dogmatic religious conservatives among the Dominican order); a man involved with artisans and academic humanists alike; an excellent artist who produced his own plates for several hundred printed works; architect of one of the most impressive Renaissance utopias; educational reformer; a Lutheran theologian and social reformer who saw the extension of scientific learning as crucial to societal advance; an advocate of alchemy and allegorical knowledge, as well as of the mundane application of knowledge in Christian charity; and a man who quite literally mediated between the magical emphasis of Campanella and the technological emphasis of Bacon. We look briefly here at the life and works of this man, Johann Andreae (1586–1654), to get some sense of how all of these traditions came together at the end of the Renaissance and the beginning of the seventeenth century.[1]

THE FUSION OF RENAISSANCE TRADITIONS IN THE LIFE OF JOHANN ANDREAE

Johann Valentin Andreae was born in 1586 at Herrenberg, near the Black Forest in present day Germany, fifth son of a Lutheran pastor and grandson of one of the central authors of the *Formula of Concord*, which established official Lutheran church doctrine. His mother, born Maria Moser in 1550, was interested in medicine and drugs; and when Andreae's father died in his mid-forties, she not only supported the family as a pharmacist but even took the position of court apothecary. Throughout most of Andreae's youth medical students boarded with the family; from them Johann learned of both Galenic and Paracelsan medicine.

At age 15 Johann went to Tubingen to study theology. In 1603 he received his bachelor's degree and two years later he became a master of arts. He studied Lutheran orthodox theology, but also Greek and Latin, and mathematics and astronomy from Michael Maestlin (1550–1631), an outspoken Copernican and the principal teacher of Johannes Kepler. Maestlin's influence on Andreae was substantial. Through him Andreae was introduced to the technical details of Copernican astronomy and to Kepler, with whom Andreae carried on a regular correspondence until Kepler's death. Another lifelong friend of Kepler, the law professor Christopher Besold (1577–1638), opened his personal library—which emphasized alchemical, astrological, and other occult works—to the young Andreae. As a result, Johann became steeped in Hermetic literature. In 1605 he wrote a tract, paralleling Pico della Mirandola's attack, on false and popular astrology in the name of true astrological and astronomical knowledge.

In the same year Andreae began work on one of his two most important writings, *The Hermetic Romance: Or the Chemical Wedding of Christian Rosenkrantz*, which was published in 1616. In this alchemical allegory, popular alchemists and gold makers are attacked in the name of a true Christian alchemy, and the seven supposed stages of the alchemical "work"—distillation, calcination, putrefaction, solution-dissolution, coagulation, vivification, and multiplication-projection—are used as allegorical representations of the stages of a Christian's search for (and final achievement of) salvation in union or marriage with Christ.[2] Much to Andreae's surprise, when *The Chemical Wedding* appeared it was taken by many as a central document of the secret Rosicrucian movement,[3] and its popularity brought both fame and notoriety to the author.

Andreae had intended to stay at Tubingen until ordination, but in 1606 he was asked to leave because of his (probably innocent) association with a group of students who were involved with local prostitutes. After two years of wandering throughout Germany, Andreae returned to Tubingen where he served as tutor to two young aristocrats. At the same time, he followed Juan Luis Vives' advice regarding manual knowledge by arranging to study with a local clockmaker, a goldsmith, and a printer. By 1610 Andreae was allowed to resume his theological studies at Tubingen under Andreaus Osiander, son of the man who had undertaken the publication of Copernicus' work and had written the careful "letter to the reader."

During the next couple of years Andreae wrote a small work, *De Christiani Cosmoxeni genitura (The Christian Wandering in the Universe)*, in which he began to explore the value of scientific learning for what he termed the marriage of sanctified faith with sanctified life.[4] The true Christian is "God's legate to men, their moral critic and guide, their physician and example, and an architect of society . . . He is a king—not only of the 'microcosm' (himself) but also of the 'macrocosm' . . ."[5] In order to carry out his mandate from God, he is given servants, "The Angels, the whole of nature, all true philosophy, arts and sciences."[6] Throughout the remainder of his life Andreae continued to reiterate this theme, which contained the essence of the Hermetic message without the Hermetic emphasis on the occult.

Later, Andreae expressed his version of this doctrine in a powerful passage that invites us to recall the *Pimander* of Hermes Trismagistus. In *Civis Christianus (The City of Christ)*, printed ca. 1619, Andreae again presents us with the Christian wanderer seeking meaning for his life. Totally exhausted, the wanderer in despair turns to God.

It was fitting to ask and listen to Him who had made and enriched our little world with such wisdom that it represents to the life the great world, and possesses no fewer wonders than the great world itself. For man is a creation of the Almighty and displays in miniature not only the movements, forces, revolutions, sympathies, and antipathies of the heavens and the earth, but also the reconciling advent of the Christ. How sad it is that thousands of men roam the earth blind and deaf to this knowledge of themselves!

While even yet staggered and trembling from head to toe as a result of Satan's attacks, I heard a penetrating, vivifying voice crying out behind me: 'Rise up!' which, as it were, lifted me immediately to my feet and infused me with new strength. Looking all about me, I heard it cry again: 'Come!' But as soon as I hesitated, it said, 'Go Back!', and it turned me completely in and out, or, more accurately, it turned me inside myself.

Then in the half light there appeared a very small and humble sanctuary, having but a single portal, and to which only a darkling illumination was admitted through slits. Entering within, my sight at first failed me, but then came back little by little. Whereupon I saw various amateurish statues and paintings representing virtues, and they were half eaten away. Among others, there were Wisdom, Justice, Temperance, Courage, Prudence, and the like, mutilated and broken for the most part, as if barbarians had passed by. Then I saw, lying here and there, the demolished wheels, gears, and other parts of an armillary sphere—fragments of an ingenious piece of work, which no master craftsman would be able to put back together. Also I perceived deplumed wings, broken ladders, and abandoned blocks and pulleys—vestiges of some great but vain project. I found all in confusion and no one to tell me what it signified.

O the ineffable radiance that suddenly appeared! A light descended through the vault of this chapel—a light in human form, with flesh and body in every way like us, yet in splendor it was unquestionably God. His resplendence so illumined the place that what was heretofore hidden now stood revealed. The images of the virtues, with marvellous artifice, reflected the light like purest crystal and even began to move as if alive; now they were whole and without any defect. And the parts of the sphere came together to form a most exquisite timepiece, which portrayed with stupendous accuracy the movements of the universe and the plan [*idea*] of God's government. Everything displayed perfection.

The God-man, dressed in a white robe, carried a book and a cross to establish the rule of holy theory and practice. This rule suits man so well that though the book equals all the world's ordinances and the cross all the world's burdens, even a child can carry them in its two hands. I perceived that the God-man stood opposite the world, and the world opposite Him. Then He gently spoke to me: 'My son and brother, you have at length come to the point of practicing what I have taught you for so long. People innumerable make profession of knowing Me; very few prove it in fact. There are many Christians, but few Christ-bearers; many carriers of the Gospel, few carriers of the Cross. The world convinces itself that it is enough to know my will intellectually. As for you, I embrace you, kiss you, make you mine, so that afterwards, when you see Me and hear Me speaking, you will receive what I teach and imitate what I do, follow my desires and commands, and remain patient when I chastise you. Know that you cannot have My book without My cross, nor My cross without My book: think not to possess My doctrine without My life and example. You can be joined to Me only by the indissoluble bonds of love.' And Christ stretched forth His right hand and cleansed me completely with the divine dew, refreshed me with the celestial manna, and commanded that I be and remain His and receive His name.

During my sojourn in the world, I thought only of the Christ with trembling. But now, in all joy and in filial confidence, I said to Him: 'Lord Jesus, my King, take me to yourself and deliver me from the world. Command what You will, and give me the grace and power to accomplish it without excuse or argument.' And as I said this, the little sanctuary changed into a magnificent temple, divided into many courts; it became a spacious tableau of the universe.[7]

In this mythic parallel to Hermes' dream confrontation with *Nous*, Andreae makes two crucial modifications in Hermetic doctrines. In the first place, he purges them of the objectionable pridefulness and arrogance of the early Renaissance versions. When Andreae's Christian looks within he sees only the husks of virtues and the broken remnants of human attempts to understand and modify the world—until Christ enters to give life and purpose to human knowledge of the universe and to attempts to undertake projects. In this way he is able to reintegrate the Hermetic dreams of power with at least *Protestant* religious orthodoxy. Humanity shares in divinity not by its nature, but only by God's grace. In the second place, we see that when human understanding and powers are given life through Christ, they are not the magical insights and powers associated with speculative Hermeticism, but knowledge of the universe derived from mathematical astronomy (symbolized in the armillary sphere) and the powers of the mechanic, who works with ladders and blocks and pulleys.

Andreae *does* retain the emphasis on microcosmic-macrocosmic analogies from earlier humanistic and Hermetic ideas, and the absolute insistence that faith and knowledge must be united in Christian acts or works; they must not remain in the realm of mere contemplation.

In 1613 and 1614 Andreae put to work the skills he learned from the goldsmith, clockmaker, and printer a few years earlier, linking them directly to his scientific interests. In 1613 he assisted his old theology teacher, Matthias Haffenreffer, by preparing a series of maps and elegant perspective drawings for his *Templum Ezechielis*; and in the following year he published a mathematical text, *Collectianeorum mathematicorum decades XI*, that included introductions to plain and solid geometry, arithmetic, statics, astronomy, gnomonics, optics (including perspective), architecture, and mensuration. He illustrated this text with 110 of his own copper plates, including two derived from Dürer's works on perspective and the proportions of the human figure. Finally, he ended the textual introduction with the marvelous Christian-Copernican line, "Let there always be glory to the true God, omnipotent, one, the sun."[8]

During this same period Andreae was introduced by a friend to the ideas of Thomasso Campanella, and while he did not embrace Campanella's magical emphasis, he later translated some of Campanella's poetry into German. He frequently praised Campanella's fight against "tyranny, sophistry, and hypocrisy" during his own (losing) fight to keep secular powers from exerting undue influence on the Lutheran Church; and he toyed with the idea of entitling his own Utopian work *Civitas Soli*, a title clearly associated with the Calabrian Revolt spearheaded by Campanella.

In 1614 Andreae was finally ordained as a Lutheran pastor. He continued his literary output (he wrote about 600 works before he died in 1654), but he also began to put into practice his theory of Christian works. On the local level he organized a mutual aid society, the "*Christich Gottliebende Gesellschaft*" (Christly, Godloving Society), among the clothworkers at Calw. This society, which began with thirteen families in 1621, undertook to provide education and social services, especially to widowed or orphaned dependents of members. A major forerunner of later aid societies, this one still remains active today.[9]

On a much larger scale Andreae instituted plans for a "*Societas Christiana*" (Christian Society) of learned men who would labor for the spiritual, educational, material, and moral betterment of society. Though we know the names of some of the early members of this society, little was known of its general scope and influence until 1944 when two of its founding documents were discovered among the papers of Samuel Hartlib, a leader of one of the groups whose merger eventually produced the Royal Society of London. So far as we know, Andreae's society never came into existence in the form he envisioned; but some passages from John Hall's English translation of Andreae's work, *A Modell of a Christian Society*, give us a good idea of his aims and of the role of science in their attainment. The chief goal was,

> After confession of the reformed religion which we call Evangelical or Protestant ["Lutheran" in the original], and after profession of reformation of life, to follow all means of knowing truth and doing good, to make use of what is already found out, [and] to discover things falsely reputed for truths . . .

In order to attain these goals Andreae proposed a "Collegium" which should embrace, among others, naturalists, including "Physicians, Anatomists, [and] Surgeons"; mathematicians, including "musicians [and] Mechanicks"; and philosophers, including "Chymists [and] Metallists."[10]

As Felix Held claimed long before these documents came to light, there can be little doubt that through Hartlib, Hall, John Drury, Robert Boyle and Jacob Commenius (the famous continental educational reformer whose ideas were also at least partially derived from Andreae's works), Andreae's ideas about a Christian society played a significant role among those who later founded the Royal Society.[11]

ANDREAE'S *CHRISTIANOPOLIS*: THE CULMINATION OF CHRISTIAN HUMANIST SCIENTISM

In order to get an idea of precisely how Andreae thought that the investigations of "Chymists, Metallists, Mathematicians and Mechanicks" could lead to the betterment of mankind we must turn to his most famous work, the utopian *Christianopolis*, published in 1619. This work expresses the culmination of Andreae's belief in the benefits that scientific investigation might bring to Christian society—*when properly subordinated to genuinely religious aims.*[12]

Once again Andreae's Christian wanders through the world.

> While wandering as a stranger upon the earth, suffering much in patience from tyranny, sophistry, and hypocrisy, seeking a man, and not finding what I sought, I decided to launch out once more upon the Academic Sea, though the latter had often been very hurtful to me. And so ascending the good ship Phantasy, I left the port together with many others and exposed my life and person to the thousand dangers that go with desire for knowledge.[13]

This time, like Thomas More's hero, Raphael Hythloday, Andreae's is shipwrecked and washed up alone on a romantic island, the site of the perfect city, Christianopolis. He is interrogated by his hosts to ensure that he is not some kind of deviate like the "drug mixers who ruin the science of chemistry,"[14] that he is a pious Christian, and that he has made some progress in such things as "the observations of the heavens and the earth, in the close examination of nature, in instruments of the arts. . . ."[15] Then he is turned loose upon the city with three guides.

The economic arrangements of Christianopolis reflect More's pure Christian communism. The governmental structure is that of theocratic Geneva. And the theology is that of orthodox Lutheranism. Of these, we need say little. Rather we must focus on the nature of industrial activity, education, and their interrelationships. Though Andreae emphasizes that agricultural activity is thoroughly grounded in a knowledge of soil fertility and animal biology, the first major objects of his attention are the metallurgical workshops. Here, he says,

> in truth you see a testing of nature herself; everything that the earth contains in her bowels is subjected to the laws and instruments of

science. The men are not driven to work with which they are unfamiliar, like pack animals to their task, but they have been trained long before in accurate knowledge of scientific matters, and feel their delight in the inner parts of nature. If a person does not here listen to the reason and look into the most minute elements of the macrocosm, they think that nothing has been proved. Unless you analyze matter by experiment, unless you improve the deficiencies of knowledge by more capable instruments, you are worthless. . . . Here one may welcome and listen to true and genuine chemistry, free and active; whereas in other places false chemistry steals upon and imposes upon one behind one's back. For true chemistry is accustomed to examine the work to assist with all sorts of tests, and to make use of experiments. Or to be brief, *here is practical science.*[16]

Turning to other manual arts, he praises Christianopolan clock makers, organ makers, cabinet makers, sculptors and masons, pointing out that such men in Christianopolis are all educated men.

Within the central square of the city lie the centers of religion, government, and learning. Here at the focus of city life lie a series of laboratories. One is devoted to chemistry.

Here the properties of metals, minerals, and vegetables and even the life of animals are examined, purified, increased, and united for the use of the human race and in the interests of health; . . . here men learn to regulate fire, make use of the air, value the water, and test earth.[17]

The second is a pharmacy. A third is an anatomy theater where dissections are done so that men may learn to "assist the struggles of nature."[18] Next there is a natural history museum; and that is followed by an artist's workshop in which paintings are created and "architecture, perspective, methods of pitching and fortifying camps,"[19] are specifically taught. Here, too, those who will not be directly involved in art, architecture, and engineering are taught artistic judgment and aesthetic appreciation. Finally, there are two mathematics laboratories, one filled with astronomical instruments (including the recently discovered telescope), and a second that includes star charts along with models of tools and machines.

Andreae soon launches into an extended discussion of education that covers over a third of *Christianopolis.* Christianopolan education is, of course, heavily weighted toward theology; but it includes a secondary emphasis on the sciences, beginning with arithmetic and geometry, and moving through astronomy and astrology, natural philosophy, and medicine. The general thrust of Andreae's discussion of scientific learning can be gleaned from his comments on geometry, astrology, and natural philosophy. Of geometry Andreae says,

> it adapts itself especially to human wants and applies the deepest propositions and theorems to practical matters with admirable diligence. . . . If one desires theoretical research, nothing is more subtle; if one desires to apply it to practical problems, nothing is more convenient or rapid. If you entrust to it any talent, the same is returned nimble and applicable to anything. Hence the inhabitants of Christianopolis set much store by it, since they see that there is no art that is not rendered easier by it.[20]

One could not want a clearer presentation of the nonmystical attitude to useful mathematics.

Andreae's attitude toward astrology is fascinating. The effects of the sun and moon, he says, are easily seen, but of the effects of other stars there are many conflicting opinions. In general, Christianopolans do not believe in any specific horoscopic system. "And so they emphasize rather this, as to how they may rule the stars and by faith, shake off the yoke if any exists."[21] To him who denies any validity to astrological influences, however, Andreae offers one of his most ironic comments: "I would wish that he would have to dig in the earth, cultivate and work fields, for as long a time as possible, in unfavorable weather."[22]

Andreae's attitudes toward mathematics, chemistry, and even astrology are almost exclusively applied in character. That is, one seeks to know *only* in order to do. But in connection with astronomy to some extent and natural philosophy even more, he draws from the older tradition of natural theology and fuses it with the new emphasis on utility. Through natural philosophy,

> we discover of what material things are made, what is their form, measure, place, and time; how the heavens move and how they appear, how elements mingle and how they increase, for what purpose living animals and plants exist, of what use metals are, and especially, what the soul, that spark of divinity within us, accomplishes. All these, forsooth, are very beautiful things, and it is below his dignity for man not to know them, after the faithful investigations of so many men. For we have not been sent into this world, even the most splendid theater of God, that as beasts we should merely devour the pastures of the earth; but that we might walk about observing his wonders, distributing his gifts, and valuing his works. . . .
> "It is . . . man's *duty*, now that he has all creatures for his use, to give thanks to God Himself in the place of them all; that is, he should offer to God as much obedience as he observes in His creatures. Then

he will never look upon this earth without praise to God or advantage to himself; . . . *Blessed are they who use the world and are not used by it.* . . . He who recognizes Christian liberality will never subject himself to the base servitude of creatures."[23]

In this and other passages from *Christianopolis* one can see the modern scientistic vision—the "better living through chemistry" mentality—being born out of the merger of humanistic emphases on service to mankind, the mathematicization of Renaissance art and architecture, the dreams of Paracelsan alchemists, and the substance (but not necessarily the spirit) of classical natural philosophy, mathematics, and astronomy. From this time forward, increasing numbers of men would search for, and expect to find, the sources of improvement in health and material circumstances through the investigation and exploitation of natural phenomena. Soon they would begin to seek answers to even their social and political problems through observation and quantification, rather than through revelation and traditional philosophical reflection.

In Andreae's hands the reform of life and nature were constantly subordinated to a spiritual reformation and controlled by a set of religious priorities; but in the hands of his older contemporary, Francis Bacon, and those who followed, the notion of reform was rapidly secularized and turned into the notion of improvement. Andreae's pious hope that men would learn to use nature rather than be used by it was ultimately turned into Daniel Defoe's crass doggerel justification of exploitation:

Nature's a Virgin, very chaste and coy.
    To court her's nonsense, if you will enjoy,
She must be ravish't—
    When she's forced, she's free,
A perfect prostitute to Industry.[24]

THE TRANSFORMATION OF RENAISSANCE TRADITIONS IN THE WORKS OF FRANCIS BACON

The greatest difference between Andreae's vision of the *reform* of the human condition through scientific knowledge and Francis Bacon's vision of the *improvement* of human life through science is indicated in a derogatory remark made by the great physician and anatomist, William Harvey. We can say without prejudice that Johann Andreae wrote of natural philosophy like a pious theologian. Harvey said of Bacon—with extreme prejudice, but great insight—that he "writes philosophy like a Lord Chancellor."[25] By this remark, Harvey intended to indicate that Bacon's ignorance of the tedious and painstaking details involved in actually doing scientific research had led him to an overly simplistic and optimistic view of the possibilities of his own proposed method. Harvey also clearly thought that Bacon was far too impatient and imperious in his vitriolic rejection of older traditions—like those of Aristotle—from which Harvey knew there was much valuable insight yet to be gained. But Harvey's comment contains another truth that was probably unintended. It suggests that secular power, rather than spiritual wellbeing or pure knowledge, was at the heart of Bacon's interests.

It would be wrong to deny to Bacon any religious motivation. Like the earlier Italian Humanists, and like Andreae, Bacon frequently formulated his aims in terms of Christian charity and the religious duty to serve mankind. But the emphasis in Bacon's works was different. One can sense this difference even when Bacon is most explicitly treating religious themes. In one of his first published works, the *Sacred Meditations* of 1597, Bacon reflected on the miracles of Jesus, emphasizing not Christ's role in the salvation of Christian souls, but the fact that all of his miracles were concerned with bodily rather than spiritual restoration:

> He restoreth motion to the lame, light to the blind, speech to the dumb, health to the sick, cleanness to the lepers, sound mind to them who were possessed with devils, life to the dead. There was no miracle of judgement, but all of mercy, and all upon the human body.[26]

Even if he was technically correct, to focus on material concerns instead of, rather than in addition to, spiritual ones in connection with Christ's ministry to man was unusual. And these priorities were often restated and specifically connected to Bacon's positive attitude toward the crucial significance of scientific knowledge. Thus, he wrote,

> Man by the Fall fell both from his state of innocence and from his dominion over the creation. Both of these losses can, however, even in this life be in some measure recovered, the former by religion and faith and the latter by the arts and sciences. For the creation was not by the

curse made altogether and forever rebellious to man. In virtue of that charter, 'In the sweat of thy brow thou shalt eat bread,' but assuredly not by disputations nor by vain magical ceremonies, the creation is at length and in some measure being subdued to the supplying of man with bread, that is to say, with the satisfaction of *human* needs.[27]

Note that Bacon specifically disassociates spiritual salvation from the notion of man's dominion over nature in this passage. Similarly he specifically divorces religion and faith from the arts and the sciences. For Bacon, secular knowledge is not sanctified and given meaning only by God's grace. The two are essentially independent domains. In the *New Organon* (1620) he reiterates this emphasis on the separation of religious and philosophical topics with a vehemence that reminds one of William of Ockham. Natural knowledge comes only from the application of empirical methods, but religious doctrine, "as well moral as mystical, is not to be obtained but by inspiration and revelation from God." From any "mixture of things human and Divine there arises not only a fantastical philosophy, but also a heretical religion."[28] So far are sacred and profane knowledge from one another that religious beliefs are more likely to contaminate our knowledge of the natural world than to enhance it, and vice versa.

Of course, once independent scientific knowledge is obtained, it should be utilized in a spirit of Christian charity; and Bacon is casually certain that this will be the case.

If the debasement of arts and sciences to purposes of wickedness, luxury, and the like, be made a ground of objection [to the search for scientific knowledge], let no one be moved thereby. . . . Only let the human race recover that right over nature which belongs to it by divine bequest, and let power be given it; the exercise thereof will be governed by sound reason and true religion.[29]

To be fair to Bacon we must admit that this naive expectation regarding the human exercise of power—especially naive for one who had been at the center of English court intrigues for over a quarter of a century—was tied to a personal commitment to traditional humanistic ideals of service. He condemned those who sought power over their fellow citizens, and merely tolerated those with nationalistic aims, who "labor to extend the power of their country and their dominion among men." Only those, who "endeavour to extend the power and dominion of the human race itself over the universe,"[30] received his unstinting praise and admiration. Yet when we isolate the ringing phrases in which Bacon announced his goals for scientific knowledge they seem much more strident, materialistic, and exploitative than those of any of his predecessors. He sought, "the empire of man over things,"[31] and "the enlarging of the bounds of Human Empire, to the effecting of all things possible."[32] He wanted to "increase and multiply the revenues and possessions of man," and he wanted to "bind [nature] to your service and make her your slave."[33] Bacon thus shifted the focus of man's dominion over nature from the spiritual aims that it had in the magical traditions and in the hands of Andreae to one that was largely secular and materialistic.

He also totally abandoned the older magical emphasis, which had remained in a muted form in Andreae's acceptance of "true" alchemy and astrology. Bacon saw all magical sciences as vain and imaginary. While he lauded the *aims* of natural magic, astrology, and alchemy, he deplored the practices, all of which, he said, "are full of error and vanity."[34] In one of his many shrewd insights Bacon pointed out why alchemy had had some incidental good effects in spite of its fundamental falsity.

> . . . to alchemy, this right is due, that it may be compared to the husbandman whereof Aesop makes the fable, that when he died told his sons that he had left unto them gold buried under the ground in his vineyard, and they digged over all the ground, and gold they found none, but by reason of their stirring and digging the mould about the roots of their vines they had a great vintage the year following; so assuredly the search and stir to make gold hath brought to light a great number of good and fruitful inventions and experiments, as well for the disclosing of nature as for the use of man's life.[35]

Just as Bacon built upon the Hermetic thrust toward the human dominion over nature, but successfully turned attention away from the earlier magical and religious emphases, he tried with equal vigor to transform the growing tradition of interactions between scientific activity and artisanship that had developed in connection with engineering. Like George Agricola, whose *De Re Metallica* he cited over one hundred times in his writings, and like Juan Luis Vives, whose educational ideals he imbibed from his father and his father's friends, Bacon saw in artisan activities the prime source of human progress. And he saw in the aristocratic intellectualism of Aristotle and other "philosophers" a real impediment to progress. But once again Bacon changed the emphasis of the tradition from which he borrowed. In the first place, Bacon seems to have been mathematically illiterate and incapable of recognizing the extent of the mathematical element in the mechanical arts and engineering of the Renaissance. For that reason, and because he was vehemently opposed to virtually all speculative traditions—whether associated with Aristotle, Plato, Paracelsus, Ockham, Hermes, or any other philosopher—Bacon acknowledged no value to any scientific knowledge not strictly derived from manual activity.

Once again, in fairness to Bacon, we must admit that he provided excellent analyses of the weaknesses of all current systematic philosophical thought. His discussions of the "Idols" of the human mind provide shrewd and unimpeachable insights into many of the ways men can be misled in their theorizing about nature.[36] But in his overt unwillingness to consider the possible value of prior philosophical speculation, Bacon was inclined to throw out the baby with the bath water. This is particularly obvious in connection with his attitude toward mathematical theorizing; for his distrust of Platonic and Pythagorean number mysticism carried over into a dismissal of even the applied mathematical tradition associated with Archimedes.

To a great extent Bacon's antiphilosophical biases were taken directly from the most radical antiintellectual polemics of artisans like Bernard Palissy, a Parisian potter, whose lectures advocating a veritable cult of things Bacon may have attended in his youth.[37] But while Bacon shared the artisan emphasis on empiricism, he felt that the experiments of mere craftsmen were unfocused and that the accidental discoveries and improvements they generated provided an unacceptably uncertain source of future progress.

Given his simultaneous antagonism toward the intellectual tradition of natural philosophy and his impatience with undirected empiricism, Bacon immodestly proposed a total reformulation of all knowledge.

> There was but one course left, therefore—to try the whole thing anew upon a better plan and to commence a total reconstruction of sciences, arts, and all human knowledge, raised upon the proper foundations.[38]

Like both the magical and artisan traditions from which it departed, Bacon's new tradition would be aimed at "the invention not of arguments, but of arts." It would seek not "to overcome an opponent in argument" but "to command nature in action."[39]

To attain these aims, Bacon sought to abandon the syllogistic and deductive logic of mathematics and earlier sciences and to depend solely upon induction from experience, beginning with particulars and only gradually moving toward more and more general propositions. In addition, Bacon's method would provide a new kind of induction—one not simply based on enumeration and fallible sensory experience. His method would be designed to take into account the limits of the senses and raw experience by focusing on experiments in which nature is constrained, and various conditions are controlled and varied one at a time by the investigator. Bacon's method would be one of controlled experiment, rather than one of mere experience.

It is unnecessary to follow Bacon into the details of his method—most of which were of marginal subsequent importance. It is, however, important to realize that he was without doubt the most widely read and most effective seventeenth-century polemicist for four fundamental views: (1) that human knowledge should be both aimed at and evaluated by the power that it provides to improve the material conditions of human life; (2) that only a thorough understanding of nature, as it exists independently of human beings and their wishes, can provide the foundation of useful knowledge; (3) that the proper method for gaining useful natural knowledge is grounded in experimentation, and that inadequately supported speculations must be exorcised from natural philosophy; and (4) that natural knowledge must be radically separated from religious and political attitudes—except for decisions regarding the application of independently discovered knowledge to the amelioration of the human condition.

Viewed from within the scientific specialty, Bacon's methodological writings have varied in their impact over time. Most serious scientists probably shared Harvey's unfavorable assessment when they first appeared. Since that time scientists have generally found Bacon's antagonism to hypothetico-deductive methods too constricting, though they have welcomed his positive emphasis on experimentation and on the public character of science. But for a period of about thirty years at the end of the seventeenth century in England, during most of the eighteenth century in France, and much of the nineteenth century in America and England, there were very strong Baconian movements even among natural scientists.

Regardless of their methodological significance—or insignificance—Bacon's views played a very important role, beginning in the 1640s, in generating widespread public interest in (and support for) scientific activities. Initial response to Bacon's *Great Instauration* (1620) and *New Organon* (1620) was not enthusiastic; but in 1624, he reexpressed the basic elements of his program in the form of an immensely popular utopian tract, the *New Atlantis*, which was published in an unfinished form after his death, in 1627. In the depiction of Salomon's House in *New Atlantis* we find the most powerful and influential expression of seventeenth-century scientism, linked to a resounding call for the public support of scientific research in the service of human material well being.

BACON'S *NEW ATLANTIS*: THE BIRTH OF MODERN SECULAR SCIENTISM

In his systematic writings on science up to 1624, Bacon had clearly been guilty of several self-defeating procedures. In the first place, even in an era that tolerated overt egoism in print, Bacon had alienated readers by praising himself and railing against virtually all others—whether intellectuals or artisans. Second, he had almost certainly frightened those with the political power to help initiate his programs by hinting broadly that his methods would have democratic, or at least republican political and economic implications. He specifically argued that one of their consequences would be to make England more like the Low Countries, where "wealth is dispersed in many hands and not ingrossed into a few." Similarly, he expected use of his methods to concentrate wealth "in those hands where it is likest to be the greatest sparing and increase, and not in those hands wherein there useth to be the greatest expense and consumption."[40] Finally, he had worried figures of at least the moderate and conservative portions of the religious spectrum because his emphasis on the worldly aspects of religion seemed to link him with radical millenarianism. His total divorce between religious and natural knowledge might have pleased conservative Anglicans, but his emphasis on works and his de-emphasis of spiritual concerns alienated them. His emphasis on works was thoroughly compatible with moderate Protestantism, but his abandonment of natural theology was inconsistent with moderate reform attitudes on the continent and in England.

In the *New Atlantis*, Bacon managed to avoid almost all of the potentially alienating characteristics of his earlier writings without explicitly drawing back from any of his arguments. Thus he was

able to make his ideas palatable to political and religious conservatives without driving away his more radical following. By putting his
own views in the mouth of the fatherly and pious director of Salomon's House, he could avoid the abrasive sense of arrogance and
self-congratulation that informed his other writings. And by placing
his idealized scientific institutions in a society that was portrayed
as overwhelmingly devoted to Christian spirituality, was organized through elaborate social hierarchies, and was characterized
by what one can only call *conspicuous consumption*, he could
implicitly deny that his ideas had any dangerous religious or social
consequences.

From this perspective there is something faintly distasteful and
hypocritical about Bacon's portrayal of experimental science as the
dominant institutional feature of a pious and authoritarian society;
but one cannot quarrel with his rhetorical success. By masking any
uncomfortable implications of the ideals it espoused, the *New
Atlantis* found admiring readers from all social classes and from all
religious sects.

The book begins in the standard formulaic way for a Renaissance utopia. The narrator's ship is blown off course in a monstrous
storm. After a period of wandering in despair, the crew prays for
salvation, and they sight the shores of a marvelous uncharted island.
But Bacon signals a new direction immediately; instead of the usual
wise and well-fed, but humble and modestly attired, resident passing by to greet the ship, Bacon offers his survivors a very different
introduction from

> a person (as it seemed) of place. He had on him a gown with wide
> sleeves, of a kind of water chamolet [a costly, finely woven camel's hair
> fabric], of an excellent azure colour, far more glossy than ours; his
> under apparel was green, and so was his hat, being in the form of a
> turban, daintily made, and not so huge as the Turkish turbans, and the
> locks of his hair came down below the brims of it. . . . He came in a
> boat, gilt [i.e., covered with gold leaf] in some of it . . .[41]

We thus see from the outset that the New Atlantis is to be a much
grander and wealthier place than More's Utopia, Campanella's City
of the Sun, or Andreae's Christianopolis. There will be no renunciation here of the opulence that is the due of social position.

Nor are we left long in doubt about the religious orientation of
the island. The first words spoken by the official ambassador who
greets the ship are, "Are ye Christians?"[42] When the shipwrecked

sailors begin to ask questions in turn, *their* first questions are directed at how such a remote place came to be converted to the Christian religion. Before he gives the answer, their host replies, "Ye knit my heart to you by asking that question in the first place, for it showeth that you *first seek the kingdom of heaven.*"[43] In case anyone thinks that Bacon was not completely aware of the impression left by such an outpouring of religious emphasis, he or she need only reread the last sentence of Bacon's statement.

It would be tedious to follow Bacon's repeated allusions to the great wealth, opulence, and spirituality of the people of Bensalem (the inhabitants' name for the island). Nor is it necessary to describe their ornate ceremonies or their laws and governmental institutions, which are at once rational and monarchical. Instead we move directly to the final long speech of the work in which the director of Salomon's House speaks to the visitor about the state of this institution. We can do this without seriously misdirecting any potential reader, for as Bacon's literary executor pointed out in publishing *New Atlantis*, the whole fable was devised as a setting for presenting "a model or description of a college instituted for the interpreting of nature and the producing of great and marvelous works for the benefit of men, under the name of Salomon's House."[44]

For one last time, even from the grave, Bacon restates his great ambition for mankind:

> The end of our Foundation is the knowledge of causes and secret motions of things, and the enlarging of the bounds of Human Empire, to the effecting of all things possible.[45]

The society of *New Atlantis* may admire religious spirituality, but the aims of Salomon's House are conspicuous in their lack of religious reference. It is *Human* Empire that its practitioners seek to expand. And in order to expand human empire they seek first to develop an organized, universally valid, and testable body of knowledge about natural phenomena based upon a system of concepts, rules of procedure, and model investigations that are shared by a well-defined group of practitioners. They seek to create a *scientific* basis for material progress.

The scale of the scientific facilities and activities of Salomon's House are suited to the grandeur of New Atlantan society and of Bacon's ambitions. For underground experiments and refrigeration there are artificial caves sunk six hundred fathoms beneath the surface of the earth; and for meteorological observations and experi-

ments there are towers which stand half a mile above the mountain-
tops and three miles above sea level. There are natural and artificial
lakes, ponds, rivers, wells, and fountains for doing experiments that
require water and water power. There are huge buildings for creating
artificial atmospheric effects; a variety of mineral baths for establishing the medical effects of different solutions on different conditions; orchards and gardens where tests are done on methods of
fertilizing, grafting, and what we would call hybridization; and
zoological gardens where animals are grown for experimental dissections and where breeding experiments are carried out. There are
"brew houses, bake houses, and kitchens, where are made divers
drinks, breads, and meats";[46] medical dispensaries; shops for producing new papers, cloths, and dyes; furnaces for doing chemical
experiments; "perspective houses" for optical experiments; "sound
houses" for acoustical experiments; "engine houses" for providing
new and powerful devices to move things through the air and water,
as well as on the ground; a "mathematics house" for mathematical
and astronomical studies; and "houses of the deceits of the senses"
where suspected frauds of all kinds are investigated.[47] Finally there
are two magnificent galleries where models of inventions are displayed and where statues of inventors are erected to honor those who
have been greater benefactors of mankind than any of the admirals,
generals, and statesmen traditionally honored in this way.[48]

To carry out and supervise the activities of all of these enter-
prises there is a veritable army of male and female workers and
apprentices capped by a group of thirty-six major officers. Of these,
some are specifically charged with bringing back books and information from foreign countries, some record the experiments reported
in books, some review experiments with a view to putting the reported information to work, others design and carry out new experimental programs or relate experiments already done to one another,
and some draw general principles from the work of the rest.

All in all, Salomon's House represents the first full-blown
vision of large-scale, state-supported scientific activity.

SUMMARY: THE LEGACY OF ANDREAE AND BACON

There was nothing unique about any of the specific elements that
went into creating the scientistic visions of Andreae and Bacon at the
beginning of the seventeenth century. Neither man added signifi-

cantly to the store of scientific knowledge nor to the array of methods available to be used by individual scientists, although Bacon certainly felt that he had done so. Both men were critically dependent on the magical and engineering or artisan traditions that had been in existence at least since the middle of the fifteenth century; though both turned away from the Hermetic emphasis on magic, which had failed to deliver on its early promises.

Nonetheless, Andreae's *Christianopolis* and Bacon's *New Atlantis* signaled a significant turning point in Western culture. They were important vehicles in the transference of attitudes and values that had been associated with specialist traditions among technologists and scientific intellectuals into the dominant attitudes and values of European societies. Within a few decades of the appearance of the *New Atlantis*, English country gentlemen were routinely devoting parts of their farms to experiments involving new methods of fertilizing, new crops, and new techniques of crop rotation; and they were writing to one another about their attempts.[49] The educational debates of Parliament during the period of the English Civil Wars were filled with suggestions for Baconian reform.[50] Radical Christian Communists like Gerard Winstanley were calling for the creation of a society in which empirical science was a central organizational feature; and Jean Colbert, Louis XIV's tight-fisted financial minister, had begun to provide state support to establish a scientific academy with government-salaried academicians.[51]

When Bacon and Andreae wrote, their visions were still prophetic, founded more in hope than in any experience of material progress grounded in self-consciously scientific activity. Yet, at least in part because of their powerfully persuasive polemics, this hope supported some of the most vital institutions of European intellectual life during the seventeenth and eighteenth centuries. In reformed and technically oriented educational institutions like the *Realschule*, and new universities established throughout Germany on the basis of Jacob Commenius's and August Herman Franke's urgings; in innumerable private and provincial philosophical societies founded throughout Europe; and in such prestigious institutions as the Royal College of Physicians, the Royal Society of London, and the Parisian Academie des Sciences, we see responses to the visions of Andreae, Bacon, and their readers. Of course, such institutions would almost certainly have emerged gradually even without Andreae and Bacon; for these men served primarily as catalysts who accelerated, but did not create, the importance of ongoing traditions that merged scientific and technological concerns.

The anticipated practical payoff of all of this activity was slow in coming. Most technological advance, to the extent that it was grounded in scientific knowledge at all, remained linked to elementary classical mathematics and mechanics until the middle of the nineteenth century. And this fact led to substantial impatience and disillusionment with science and scientific "projectors." Samuel Johnson was especially well aware that much opposition to science in the late seventeenth and early eighteenth centuries rose out of disappointed rising expectations produced by Baconian promises. Thus he wrote in *Idler* #88 (1759),

> When the philosophers of the last age were first congregated into the Royal Society, great expectations were raised of the sudden progress of the useful arts; the time was supposed to be near, when engines should turn by perpetual motion, and health be secured by the universal medicine; when learning should be facilitated by a real character, and commerce extended by ships which could reach their ports in defiance of the tempest.
>
> But improvement is naturally slow. The society met and parted without any visible diminution of the miseries of life. The gout and stone were still painful, the ground that was not plowed brought no harvest, and neither oranges nor grapes would grow in the hawthorn. At last, those who were disappointed began to be angry . . . and . . . the philosophers felt with great sensibility the unwelcome importunities of those who were daily asking, "What have ye done?"
>
> The truth is, that little had been done compared with what fame had suffered to promise . . .[52]

But by the second half of the nineteenth century Emile Du Bois Reymond could look *backward* rather than *forward* to a radical transformation of European life based on the work of Andreae's "Metallists and Chymists"—the applied scientists:

> Wherever something to be physically achieved remains withheld from man, there the calculus of his spirit presses forward with its magic key. . . . What the wishing wand did by magic, geology performs: freely it brings forth water, salt, coal, petroleum. The number of metals continues to increase, and though as yet chemistry has not found the philosopher's stone, tomorrow perhaps it will. For the present it vies with organic nature in the production of the useful and the pleasing. From the black stinking heaps of refuse which turn every city into a Baku, it borrows the colors by comparison with which the magnificence of tropical feathers turns pale [Reymond is speaking of coal tar dyes, on which much of the financial strength of late 19th century Germany depended]. It prepared perfumes without sun or flowers. Has it not solved Simson's riddle, how to make sweet things out of loath-

some material [i.e., saccharine out of tars]?

Gay-Lussac's preserving art has not merely wiped out the difference between the seasons on rich men's tables. . . . The scourges of smallpox, plauge, scurvy are under control. Lister's bandage protects the wounds of soldiers from the entrance of deadly germs. Chloral spreads the wings of God's sleep over the soul of pain, indeed chloroform laughs, if we wish, at the biblical curse of womanhood . . .

. . . *All peoples of Europe, of the Old and the New Worlds, travel along this road.*[53]

Johann Andreae would have admired the changes so far mentioned by Reymond; but he would have shuddered (and we must at least pause) at the attitudes that accompanied them. For Reymond saw in the power of science reason to minimize other aspects of culture—especially artistic productions and, "the one-sided ethical view of men," which he associated with the Jews.

If there is one criterion which for us indicates the progress of humanity, it is much more the level attained of power over nature. . . . Only in scientific research and power over nature is there no stagnation; knowledge grows steadily, the shaping strength develops unceasingly. Here alone each generation stands on the preceding one's shoulders. Here alone no *ne plus ultra* of the schools intervenes, no dictum of authority oppresses, while even mediocrity finds an honorable place, if it only seek the truth diligently and sincerely.[54]

The humane and sacred scientism of the Renaissance had gradually been transformed into the "objective" antihumanistic and amoral scientism of Defoe and Reymond. The process by which this occurred must be deferred to Volume II.

Notes

CHAPTER 1

1. In *The Two Cultures: and A Second Look* (New York: Cambridge University Press, 1963), Snow acknowledges that especially in the U.S. there seems to be a third emergent and isolated "culture" associated with social sciences.

2. See C. R. Rogers and B. F. Skinner, "Some Issues Concerning the Control of Human Behavior: A Symposium," *Science* 124 (1956), 1057–1064.

3. Jacques Monod, "On Values in an Age of Science." In *The Place of Values in a World of Facts: The Fourteenth Nobel Symposium* (Stockholm: Swedish Academy of Sciences, 1964).

4. Ibid.; see also J. Monod's *Chance and Necessity* (New York: Knopf, 1971).

5. See Herbert Marcuse, *One Dimensional Man* (Boston: Beacon Press, 1964), pp. 158–159.

6. See Herbert Marcuse, "Remarks on a Redefinition of Culture." In Gerald Holton, ed., *Science and Culture* (Boston: Houghton Mifflin, 1965), pp. 222–223.

7. Jacques Ellul, *The Technological Society* (New York: Vintage Books, 1964), p. xxxi.

8. See Jacques Barzun, *Science: The Glorious Entertainment* (New York: Harper and Row, 1964), pp. 9–30.

9. Quoted in Bernard Dixon, *What Is Science For* (London: William Collins, 1973), p. 12.

10. See survey results from U.S. and France reviewed by Edward Shils, "The Public Understanding of Science," *Minerva* 12 (1974), 153–156.

11. Henri Poincaré, *The Value of Science* (1913; reprint ed., New York: Dover, 1958), p. 5.

12. Ibid., p. 8. Emphasis mine.

13. Aristotle *Metaphysics* 981a.

14. See Loren R. Graham, *Science and Philosophy in the Soviet Union* (New York: Vintage Books, 1974) for a clear discussion of dialectical materialist science.

15. Christian Huygens, *Treatise on Light* (1690; reprint ed., New York: Dover, 1962), p. 3.

16. David Hume, *Enquiries Concerning Human Understanding and Morals* (1777; reprint of 3d. ed., Cambridge, England: Cambridge University Press, 1902), p. 165.

CHAPTER 2

1. James Mellaart, *Earliest Civilizations of the Near East* (New York: McGraw-Hill, 1965), pp. 44, 81–85.

2. See Samuel Noah Kramer, *The Sumerians: Their History, Culture and Character* (Chicago: University of Chicago Press, 1963), for greater detail.

3. The hydraulic theory of urbanization and social organization was stated in its strongest form by Karl Wittfogel, *Oriental Despotism: A Comparative Study of Social Power* (New Haven: Yale Univ. Press, 1957). The alternative ceremonial center thesis was effectively presented in Robert Adams, *The Evolution of Urban Society: Early Mesopotamia and Prehispanic Mexico* (Chicago: Aldine, 1965).

4. Though according to theory all land belonged to the cities' deities through the temples, it seems that from the time of the earliest records some private land ownership existed side by side with the dominant communal ownership. See Kramer, *The Sumerians*, pp. 73–78.

5. Jacquetta Hawkes and Leonard Woolley, *Prehistory and the Beginnings of Civilization* (New York: Harper and Row, 1963), pp. 659–660.

6. Kramer, *The Sumerians*, p. 229.

7. Ibid., pp. 241–242.

8. M. E. L. Mallowan, "Civilized Life Begins: Mesopotamia and Iran." In Stuart Piggott, ed., *The Dawn of Civilization* (New York: McGraw-Hill, 1961), p. 90.

9. Hawkes and Woolley, *Prehistory*, p. 668.

10. Mallowan, *Civilized Life*, p. 90.

11. See especially Cyrus H. Gordon, *The Common Background of Greek and Hebrew Civilizations* (New York: W. W. Norton, 1965) on this topic.

12. Cited in Morris R. Cohen and I. E. Drabkin, *A Sourcebook in Greek Science* (Cambridge: Harvard University Press, 1966), p. 34 n. 1.

13. Cited in Otto Neugebaur, *The Exact Sciences in Antiquity* (New York: Harper Torchbooks, 1962), p. 79.

14. Aristotle *Metaphysics* 981b 20–25.

15. B. L. Van der Waerden, *Science Awakening* (New York: Wiley, 1963), 1:78–79.

16. Van der Waerden, *Science Awakening*, p. 63 ff.

17. Cohen and Drabkin, *A Source Book*, pp. 33–34.

18. Cited in George Sarton, *A History of Science* (Cambridge: Harvard University Press, 1952), 1:37.

19. Henri Frankfort, H. A. Frankfort, John A. Wilson, and Thorkild Jacobsen, *Before Philosophy: The Intellectual Adventure of Ancient Man* (Harmondsworth, England: Penguin Books, 1949), pp. 12–14.

20. Ibid., p. 143.

21. Ibid., p. 161.

22. Cited in Stephen Toulmin and June Goodfield, *The Fabric of the Heavens* (New York: Harper Torchbooks, 1961), p. 43.

23. A. Pannekoek, *A History of Astronomy* (London: George Allen and Unwin, 1961), p. 38.

24. Samuel Noah Kramer, *History Begins at Sumer* (Garden City, New York: Doubleday Anchor Books, 1959), p. 72.

25. Kramer, *The Sumerians*, p. 227.

26. B. L. Van der Waerden, *Science Awakening II: The Birth of Astronomy* (Leyden: Noordhoff International Publishing; and New York: Oxford University Press, 1974), 2:58.

27. Ibid., p. 59.

28. Ibid., p. 48.

29. Ibid., pp. 50, 55.

30. Cited by Franz Cumont, *Astrology and Religion among the Greeks and Romans* (1912; reprint ed., New York: Dover, 1960), pp. 14–15.

31. Pannekoek, *A History of Astronomy*, p. 25.

32. Stephen H. Langdon, *Babylonian Menologies and Semitic Calendars* (London: Oxford University Press, 1935), pp. 3–9.

33. Pannekoek, *A History of Astronomy*, p. 31.

34. See R. A. Parker and Waldo Dubberstein, *Babylonian Chronology: 626 B.C.–A.D. 45* (Chicago: University of Chicago Press, 1942).

35. The most complete discussion of the schematic calendar of Selucid times appears in Otto Neugebauer, *Astronomical Cuneiform Texts*, 3 vols. (London: Lund Humphries, for Institute for Advanced Studies, Princeton, N.J., 1955).

36. Hans Jonas, *The Gnostic Religion*, 2d ed. (Boston: Beacon Press, 1963), p. 16.

37. Frederick H. Cramer, *Astrology in Roman Law and Politics* (Philadelphia: American Philosophical Society, 1954), p. 11.

38. See A. J. Sachs, "Babylonian Observational Astronomy." In the British Academy, *The Place of Astronomy in the Ancient World* (Oxford: Oxford University Press, 1974).

39. W. W. Tarn, *Hellenistic Civilization*, 3d ed. (Cleveland: World Publishing, 1952), pp. 345–346.

40. Cramer, *Astrology in Roman Law*, p. 11. Also see A. Bouche-Leclerqué, *L'Astrologie Grecque* (Paris: E. Leroux, 1899), p. 368 n. 1.

41. Van der Waerden, *Science Awakening II*, p. 162.

42. Ibid., p. 162.

43. Ibid., p. 144.

44. Plato, *Republic*, 615–620.

45. Cited in Cramer, *Astrology in Roman Law*, p. 19.

46. Theodore Roszak, "The Monster and the Titan: Science, Knowledge, and Gnosis," *Daedalus* (Summer, 1974), pp. 17–32.

47. Jonas, *The Gnostic Religion*, pp. 52, 66–67.

48. Van der Waerden, *Science Awakening II*, p. 144.

49. Jonas, *The Gnostic Religion*, p. 157.

CHAPTER 3

1. Though the ideas developed by the men of Plato's Academy and Aristotle's Lyceum, which came into existence in the fourth century, had vast importance for the Hellenistic world and beyond, they came too late to have a major impact on Hellenic Greece.

2. See M. I. Finley, *Early Greece, The Bronze and Archaic Ages* (New York: W. W. Norton, 1970), pp. 84–86, for the argument that the Homeric epics reflect more a later period of preclassical culture—from about 1200–1000 B.C. Herman Frankel, *Early Greek Poetry and Philosophy* (Oxford: Basil Blackwell, 1975), p. 44 ff, upholds the view that Homer depicts largely the situation from about 1570–1200 B.C., with the incursion of some anachronistic materials.

3. Xenophanes *Fragment* A28, in Milton C. Nahm, ed., *Selections from Early Greek Philosophy* (New York: Appleton-Century-Crofts, 1934). For a discussion of the very limited sources of our knowledge of all pre-Socratic philosophers see "The Sources for Pre-Socratic Philosophy," in G. S. Kirk and J. E. Raven, eds., *The Pre-Socratic Philosophers* (Cambridge, England: The University Press, 1957).

4. Victor Ehrenburg, *Aspects of the Ancient World* (1846; reprint ed., New York: Arno Press, 1973) p. 4.

5. See W. K. C. Guthrie, *The Greeks and Their Gods* (Boston: Beacon Press, 1950), p. 117.

6. Lowes Dickinson, *The Greek View of Life* (Ann Arbor: University of Michigan Press, 1958), p. 11.

7. Quoted in Hermann Frankel, *Early Greek Poetry*, p. 405.

8. Homer *The Iliad* 23.

9. Homer *The Odyssey* 282.

10. W. K. C. Guthrie, *The Greeks and Their Gods*, p. 39.

11. Hesiod, *The Homeric Hymns and Homerica*, trans. Hugh G. Evelyn-White (Cambridge: Harvard University Press, 1914), p. 23.

12. Quoted in Hermann Frankel, *Early Greek Poetry*, p. 478.

13. Aristotle *The Constitution of Athens* II, 1–2.

14. Frankel, *Early Greek Poetry*, pp. 255–256.

15. See, for example, Leon Robin, *Greek Thought and the Origins of the Scientific Spirit* (1928; reprint ed., New York: Russell and Russell, 1967), forward.

16. See J. Ralph Lindgren, ed., *The Early Writings of Adam Smith* (New York: Augustus M. Kelley, 1967). Though there was substantial continuity between the subject matter of ancient myth—especially Near Eastern myth—and early philosophy, there is general agreement that the Ionians, Thales, Anaximander, and Anaxemines, founded a *new* enterprise. Aristotle designated these men *Physicoi* (nature philosophers) to distinguish them from the *Theologoi* (poets of the gods) that had gone before. He did this because they focused their attention on natural phenomena more rigidly than their predecessors, and they did not appeal to the gods in the traditional way to account for these phenomena. All other domains of classical philosophy—epistemology, metaphysics, logic, political philosophy, ethics, and aesthetics—aspects of which seek to go beyond the domain of sensory experience, grew up only after natural philosophy. Moreover, they grew up primarily in response to issues raised by the natural philosophers, or as self-conscious attempts to adapt the techniques of the natural philosophers to other domains of social, political, and personal importance.

17. See Morris Cohen and Israel Drabkin, *A Sourcebook in Greek Science* (Cambridge: Harvard University Press, 1948), p. 34.

18. Ibid., p. 92.

19. G. E. R. Lloyd, "Popper versus Kirk: A Controversy in the Interpretation of Greek Science," *The British Journal for the Philosophy of Science* 18 (1967), 21–38.

20. Aristotle *Metaphysics* 985b–986a.

21. Aetius III, quoted in G. S. Kirk and J. E. Raven, *The Pre-Socratic Philosophers* (Cambridge, England: The University Press, 1957), p. 138.

22. Aristotle *Meteor* 365b6 quoted in Kirk and Raven, *Pre-Socratic Philosophers*, p. 158.

23. Hippolytus *Ref.* I8, 3–10, quoted in Kirk and Raven, *Pre-Socratic Philosophers*, p. 391.

24. Sextus *Math* 9,24 and 19, paraphrased by W. K. C. Guthrie in *A History of Greek Philosophy*, II (Cambridge: Cambridge University Press, 1965), p. 478.

25. W. K. C. Guthrie, *The Sophists* (Cambridge: Cambridge University Press, 1971), pp. 231–232.

26. Quoted in W. K. C. Guthrie, *The Sophists*, p. 228.

27. See Mario Untersteiner, *The Sophists*, trans. from the Italian by Kathleen Freeman (Oxford: Basil Blackwell, 1954), p. 4.

28. Ibid., p. 5.

29. Simplicius *Phys.* 24, 13, quoted in Kirk and Raven, *The Pre-Socratic Philosophers*, pp. 105–107.

30. Ibid., p. 114.

31. Edward Zeller, *Outlines of the History of Greek Philosophy*, 13th ed. (New York: Meridian Books, 1955), p. 46.

32. See W. K. C. Guthrie, *A History of Greek Philosophy*, I (Cambridge: Cambridge University Press, 1962), pp. 128 ff.

33. W. K. C. Guthrie, *The Greeks and Their Gods* (Boston: Beacon Press, 1950), p. 141.

34. Ibid., p. 262.

35. Ibid., p. 142.

36. Quoted in Herman Frankel, *Early Greek Poetry*, p. 331.

37. Kirk and Raven, *The Pre-Socratic Philosophers*, p. 169.

38. Ibid., p. 171.

39. Ibid., p. 168.

40. Ibid., p. 168.

41. Ibid., p. 168.

42. Ibid., p. 169.

43. Quoted by Herman Frankel, *Early Greek Poetry*, p. 333.

44. Ibid., pp. 336–337.

45. Quoted in Mario Untersteiner, *The Sophists*, p. 27.

46. See W. K. C. Guthrie, *The Sophists*, pp. 238–239.

47. Ibid., pp. 243–244.

48. Robin, *Greek Thought*, p.138.

49. W. K. C. Guthrie, *The Sophists*, pp. 108–109.

50. Ibid., p. 153.

51. Ibid., p. 151.

52. Plato *Laws* 890a.

53. Kirk and Raven, *The Pre-Socratic Philosophers*, p. 422.

54. Ibid., p. 395.

55. W. K. C. Guthrie, *The Greeks and Their Gods*, p. 340.

56. Quoted by Mario Untersteiner, *The Sophists*, pp. 43–44.

57. Plato *Theatetus* 160c ff.

58. Ibid., 167a–c.

59. Quoted in W. K. C. Guthrie, *The Sophists*, p. 195.

60. Plato *Apology* 19b–19c.

61. Aristophanes *The Clouds*, Wm. Arrowsmith trans. (New York: The New American Library, 1962), p. 34.

62. Aristophanes *The Clouds*, in Whitney J. Oates and Eugene O'Neill, Jr., eds., *The Complete Greek Drama* (New York: Random House, 1938), 2:556–557.

63. Aristophanes *The Clouds*, Arrowsmith translation, p. 47.

64. Ibid., p. 132.

65. Aristophanes *The Clouds*, in Oates and O'Neill, *Complete Greek Drama* 2:594–595, and Arrowsmith translation, p. 124.

66. Jean Jacques Rousseau, *The First and Second Discourses*, W. D. Masters, ed. (New York: St. Martin's Press, 1964), p. 50.

67. *The Clouds*, Arrowsmith translation, p. 120.

68. Aristophanes *The Frogs*, in Oates and O'Neill, *Complete Greek Drama*, 2:966.

69. Ibid., p. 962. The order of these passages has been slightly altered.

70. Ibid., p. 969.

71. Ibid., p. 973.

72. Ibid., p. 972.

73. Friedrich Nietzsche, *Complete Works* (New York: Russell and Russell, 1964), 1:89–90.

74. Ibid., p. 88.

75. Aristophanes *The Frogs* in Oates and O'Neill, *Complete Greek Drama*, 2:968–970.

76. Ibid., pp. 974–975.

77. Cedric Whitman, trans. *Aristophanes and the Comic Hero* (Cambridge: Harvard University Press, 1964), p. 246.

78. Nietzsche, *Complete Works*, 1:95.

CHAPTER 4

1. Plato *Letter VII* 325b.

2. In all comments about the evolution of Plato's thought I follow here the arguments of J. E. Raven, in *Plato's Thought in the Making* (Cambridge, England: Cambridge University Press, 1965). Experts disagree on almost every single important issue connected with Plato's ideas.

3. Aristotle *Metaphysics* 985b23–986a8.

4. Plato *Meno* 81c–82d.

5. Ibid., 81d.

6. Raven, *Plato's Thought in the Making*, pp. 69–70.

7. Plato *Meno* 86e–87b.

8. Raven, *Plato's Thought in the Making*, pp. 69–70.

9. Cited in John Herman Randall, Jr., *Plato: Dramatist of the Life of Reason* (New York: Columbia University Press, 1970), p. 245.

10. Cited in Paul Friedlander, *Plato: An Introduction* (New York: Pantheon Books, 1958), 1:94.

11. Plato *Phaedrus* 265c–266a.

12. Ibid., 266b.

13. See the introduction to almost any edition of Plato's *Timaeus*, for example, p. viii of the Bobbs-Merrill (Indianapolis) edition of 1949 with an introduction by Glen R. Morrow.

14. *Timaeus* 29.

15. The best short overall assessment of the *Timaeus* as a cosmological treatise is Gregory Vlastos, *Plato's Universe* (Seattle: University of Washington Press, 1975).

16. Plato *Laws* 889.

17. *Timaeus*, 28.

18. Ibid., 29.

19. Ibid., 29–30.

20. Ibid., 37.

21. Ibid., 41.

22. Ibid.

23. Ibid., 42.

24. Ibid., 45–47.

25. Ibid., 50.

26. Ibid., 52–53.

27. Ibid., 53–65.

28. Vlastos, *Plato's Universe*, p. 97, n. 15.

29. e. e. cummings to Eva Hesse, in F. W. Dupee and G. Stade, eds., *Selected Letters of e. e. cummings* (New York: Harcourt, 1969). p. 265.

30. Eric Havelock, *Preface to Plato* (Cambridge: Harvard University Press, 1963), p. 47.

31. Plato *Republic* 475d–476b.

32. Ibid., 395c–396e.

33. Ibid., 439d–442c.

34. Ibid., 442d.

35. Ibid., 605c–606d.

36. Ibid., 607a.

37. Aristotle *Metaphysics* 982b11–27.

38. Plato *Republic* 519c–521c.

39. Aristotle *Metaphysics* 980a21.

40. Ibid., 982b.
41. Ibid., 982a.
42. Ibid., 982a.
43. Aristotle *Nicomachean Ethics* 1102a–1103a.
44. Ibid., 1139a5–10.
45. Ibid., 1177a25–35.
46. Ibid., 1177b27–1178a 4.
47. Ibid., 1178b20–24.
48. Ibid., 25–31.
49. Ibid., 1145a6–8.
50. Aristotle *Posterior Analytics* 71b8–15.
51. Aristotle *Metaphysics* Book 13. Though Aristotle touches on the primacy of experience and abstraction from experience throughout several works, it is in *Metaphysics* 13 that independently existing ideas and mathematical entities are denied unambiguously.
52. Aristotle *Categories* 2a11–2b15.
53. See Aristotle *Posterior Analytics* 99b–110b.
54. Aristotle *Metaphysics* Book 7.
55. See especially Aristotle *On Generation and Corruption* Book II, chaps. 1–5.
56. Aristotle *De Caelo Book I*, chaps. 3, 10–12.
57. Aristotle *Categories* Books 6–8, analyse the kinds of accidents that a substance may have.
58. Aristotle *De Caelo* Book 3, chap. 2.
59. Aristotle *De Caelo* Book 1, chaps. 2–4.
60. Aristotle *Physics* Book 4, chap. 8.
61. Ibid., Book 2, chap. 3, 194b17–195a28.
62. Aristotle *Posterior Analytics* Book 1, chap. 3, 172b18–24.
63. The whole of the *Prior Analytics* is devoted to a study of syllogisms, their terms, premises, and uses. The syllogism is first formally defined at p. 246.
64. Aristotle *Posterior Analytics* 79a17–33.
65. Ibid., 71a.
66. Aristotle *Physics* 192b.
67. Ibid., 201a.
68. Ibid., 1926.
69. Aristotle *Posterior Analytics* 99b–100b.
70. See Aristotle *Topics* chaps. 13–18, for a full discussion of this critical procedure.
71. See J. E. L. Dreyer, *A History of Astronomy From Thales to Kepler* (New

York: Dover, 1953), chap. 5, pp. 108–122 for a good general discussion of the nesting spheres cosmological model.

72. Aristotle *Metaphysics* 1072a20–26.

73. Aristotle *De Caelo* 292b.

74. Aristotle *On Generation and Corruption* 336b.

75. Ibid., chap. 10.

76. See chap. 6, this volume.

77. Cited in Morris R. Cohen and I. E. Drabkin, *A Source-Book in Greek Science* (Cambridge: Harvard University Press, 1966), p. 317.

78. See, for example, G. E. R. Lloyd, *Greek Science After Aristotle* (London; Chatto and Windus, 1973), chap. 10; Desmond Lee, "Science, Philosophy, and Technology in the Greco-Roman World," *Greece and Rome* 20 (1973), 65–78; and M. I. Finley, "Technological Innovation and Economic Progress in the Ancient World," *Economic History Review* 18 (1965), 29–45.

79. Michael Avi-Yonah, *Hellenism and the East* (Ann Arbor: University Microfilms International, 1978), p. 251.

CHAPTER 5

1. See Lynn White, jr., *Medieval Technology and Social Change*, especially chap. 2, "The Agricultural Revolution of the Early Middle Ages" (Oxford: Oxford University Press, 1962).

2. John Herman Randall, Jr., *Hellenistic Ways of Deliverance and the Christian Synthesis* (New York: Columbia University Press, 1970), pp. 139–140.

3. Origin to Gregory Thaumaturgis, cited in R. B. Tollinton, *Alexandrine Teaching on the Universe* (New York: Macmillan, 1932), pp. 166–167.

4. Cited in Charles Bigg, *The Christian Platonists of Alexandria* (1968; reprint ed., Oxford: Oxford University Press, 1886), p. 241, n. 1.

5. Tollinton, *Alexandrine Teaching*, p. 117.

6. See Randall, *Hellenistic Ways*, pp. 159–160.

7. *Genesis* I:31.

8. *Psalms* 19:1.

9. See Bigg, *Christian Platonists*, pp. 196 ff. Bigg has combined and simplified elements from Clement and Origen. A complete and detailed comparison of their similarities and differences is beyond the scope of this study.

10. Cited in D. S. Wallace-Hadrill, *The Greek Patristic View of Nature* (Manchester: Manchester University Press, 1968), p. 103.

11. Cited in ibid., p. 105.

12. Bigg, *Christian Platonists*, pp. 197–198.

13. Cited in Wallace-Hadrill, *Greek Patristic View*, pp. 113–114.

14. Cited in Tollinton, *Alexandrine Teaching*, p. 142.

15. Cited in Bigg, *Christian Platonists*, p. 150.

16. Origen, cited in Wallace-Hadrill, *Greek Patristic View*, p. 120.

17. Ibid., p. 121. Emphasis mine.

18. Cited in Randall, *Hellenistic Ways*, p. 163.

19. See Nicolas Steneck, *Science and Creation in the Middle Ages: Henry of Langenstein on Genesis* (Notre Dame: University of Notre Dame Press, 1976) for a study of Henry's work and its place in Hexamaron tradition.

20. See Bigg, *Christian Platonists*, pp. 86–96, and 142–143 for a discussion of the characteristics of the Christian Gnostic.

21. Cited in Eric F. Osborn, *The Philosophy of Clement of Alexandria* (Cambridge, England: Cambridge University Press, 1957), p. 148.

22. Ibid., pp. 171–172.

23. Cited in Bigg, *Christian Platonists*, p. 49.

24. See Randall, *Hellenistic Ways*, pp. 114–117, 163, & 170–172 for a good short introduction to the Alexandrine "Logos" doctrine.

25. Bigg, *Christian Platonists*, p. 51.

26. Cited in Osborn, *Philosophy of Clement*, pp. 55–56.

27. Ibid., p. 27.

28. Ibid., pp. 25–26. Emphasis mine.

29. Ibid., pp. 25–26.

30. Ibid., p. 30.

31. Ibid., p. 134.

32. Ibid.

33. Ibid., p. 135.

34. Irenaeus, "Against Heresies," i, xiv, 5. Cited in *Ante-Nicene Fathers*, Vol. I. (New York: Scribners, 1899).

35. Cited in Henry Chadwick, *Early Christian Thought and the Classical Tradition* (New York: Oxford University Press, 1966), p. 1.

36. St. Ambrose, *Duties of the Clergy*, I, xxvi, 22. Cited in Susan C. Lawrence, " 'Nulla scientia hic florescit': Anti-Scientism in the Early Church Fathers." Unpublished manuscript in my possession, May, 1977.

37. Arnobius, *The Case Against the Pagans* (Westminster: The Newman Press, 1949), 1:11, 61.

38. Tertullian, *Ad Nationes*, ii, 4. Cited in *Ante-Nicene Fathers*, vol. 2.

39. Irenaeus, *Against Heresies*, ii, xxviii, 3.

40. Lactantius, *The Divine Institutes*, ii, vii. Cited in *Ante-Nicene Fathers*, vol. 7.

41. Irenaeus, *Against Heresies*, ii, xxvi, 1.

42. Tertullian, *On the Flesh of Christ*, 5. Cited in *Ante-Nicene Fathers*, vol. 2.

43. Cited in Gerard L. Ellsperman, *The Attitude of the Early Christian Latin Writers Toward Pagan Literature and Learning* (Washington, D.C.: the Catholic University of America Press, 1949), pp. 192–193.

44. St. Augustine, *Doctrinea Christina*, 2, 29, 46. Cited in Ellsperman, *Attitude of Early Christian*, p. 177.

45. Ellsperman, *Attitude of Early Christian*, p. 176.

46. Ibid., p. 188.

47. Ibid., p. 178.

48. Ibid., p. 185.

49. Ibid., p. 180.

50. Ibid., p. 186–187.

51. On the relations between Basil, Ambrose, and Augustine, see for example, Clarence Glacken, *Traces on the Rhodian Shore* (Berkeley and Los Angeles: University of California Press, 1967), p. 189.

52. As early as the eighth century the great English scholar, Bede, commented that Basil's work was the foundation for all works on the creation that he was familiar with. See Lynn Thorndike, Jr., *History of Magic and Experimental Science* (New York: Columbia University Press, 1923), Vol. I, p. 485.

53. St. Basil, *Exigetic Homilies* (Washington, D.C.: Catholic University of America Press, 1963), Homily 15, p. 323.

54. Ibid., Homily 1, p. 11.

55. Ibid., Homily 2, p. 25.

56. Ibid., Homily 9, p. 135.

57. Ibid., Homily 7, pp. 112–113. Emphasis mine.

58. The anonymously authored Alexandrine *Physiologus* came into existence around the beginning of the fourth century. It described the characteristics and habits of animals, drawing Christian morals from parables about each. The *Physiologus* and its medieval progeny, the *Bestiary*, were diffused more widely than any writing other than the Bible well into the twelfth century. See T. H. White, *The Bestiary* (New York: Putnam's, 1954) for a delightful translation and commentary on this literature.

59. St. Basil, *Exigetic Homilies*, Homily 9, p. 136.

60. Ibid., Homily 1, p. 3.

61. Ibid., pp. 4–5.

62. Ibid., pp. 5–6.

63. Ibid., pp. 6–7.

64. Ibid., pp. 7–8.

65. Ibid., Homily 2, p. 23.

66. Ibid., Homily 1, p. 12.
67. Ibid., Homily 2, p. 24.
68. Ibid., p. 25.
69. Ibid., Homily 1, pp. 9–10.
70. Ibid., p. 11.
71. Ibid., p. 13.
72. Ibid., p. 19.
73. Ibid., p. 14.
74. Ibid., Homily 3, p. 38.
75. Ibid., p. 40.
76. Ibid., Homily 6, p. 102.
77. Ibid., Homily 3, p. 41.
78. Ibid.
79. Ibid., Homily 5, p. 80.
80. Ibid., p. 81.
81. Ibid., Homily 3, p. 45.
82. Ibid., p. 50; Homily 4, p. 64.
83. Ibid., Homily 4, p. 60.
84. Ibid., p. 57.
85. Ibid., Homily 5, p. 67.
86. Ibid., p. 82.
87. Ibid., Homily 6, p. 90.
88. Ibid., pp. 95–101.
89. Ibid., p. 95.
90. Ibid., p. 92.
91. Ibid., p. 93.
92. Ibid., pp. 93–94.
93. Ibid., Homily 8, p. 119.
94. Ibid.
95. Ibid.
96. Ibid., Homily 3, p. 45.
97. Brian Stock, *Myth and Science in the Twelfth Century* (Princeton, N.J.: Princeton University Press, 1972), p. 240.
98. Cited in C. Warren Hollister, *Medieval Europe: A Short History*, 4th ed. (New York: Wiley, 1978), p. 19.
99. See C. W. Hollister, *Medieval Europe*, pp. 44–47, on the key role of Benedictine monasticism in medieval life.

CHAPTER 6

1. For a more detailed discussion of the role of the scientist in the medieval university than can be provided here, see Joseph Ben-David, *The Scientist's Role in Society* (Englewood Cliffs, N.J.: Prentice-Hall, 1971), pp. 45–75.

2. Edward Grant, *Physical Science in the Middle Ages* (New York: Wiley, 1971), p. 21. Emphasis mine.

3. Excellent treatments of this topic are available in Seyyed Hossein Nasr's *An Introduction to Islamic Cosmological Doctrines* (Cambridge: Harvard University Press, 1964); and *Science and Civilization in Islam* (Cambridge: Harvard University Press, 1968).

4. See David C. Lindberg, "The Transmission of Greek and Arabic Learning to the West" in D. C. Lindberg, ed., *Science in the Middle Ages* (Chicago: University of Chicago Press, 1978), pp. 52–90. The percentages come from my own categorizing and counting of texts cited by Lindberg and should not be attributed to him.

5. For an extended discussion of Abu Ma'shar's influence see R. Lemay, *Abu Ma'shar and Latin Aristotelianism in the Twelfth Century* (Beirut: Amerian University of Beirut, 1962).

6. My slightly modified version of a passage from Abu Ma'shar cited in Bryan Stock, *Myth and Science in the Twelfth Century* (Princeton, N.J.: Princeton University Press, 1972) p. 29.

7. Abu Ma'shar's sources are discussed by David Pingree in his article "Abu Ma'shar" in *The Dictionary of Scientific Biography*, 1:32–39.

8. See Brian Stock, *Myth and Science.*

9. Cited in Seyyed Hassein Nasr, *Science and Civilization in Islam* (Cambridge: Harvard University Press, 1968), pp. 210–211.

10. It is probable that the increasing frequency of translations of Aristotle, and commentaries on Aristotle, from Arabic into Latin over time in the late twelfth and the thirteenth centuries is as much a function of their increasing frequency in Islamic collections of manuscripts as of any selective emphasis by the Latin translators. One like Gerard of Cremona tended to settle in and translate what amounts to the entire curriculum in the quadruivium from some specific Arabic collection.

11. Virtually all of my comments on Averroes are derived from Part Two of Etienne Gilson's brilliant and readable *Reason and Revelation in the Middle Ages* (New York: Scribner's, 1938). Anyone interested should carefully read this book.

12. A good elementary treatment of the growth of medieval universities can be found in David Knowles, *The Evolution of Medieval Thought* (New York: Vintage Books, 1961), pp. 153–185.

13. Eventually the Masters at Paris were granted the right to admit their own members, but this came relatively late in the story.

14. Cited in Gordon Leff, *Paris and Oxford Universities in the Thirteenth and Fourteenth Centuries* (New York: Wiley, 1968), p. 139.

15. Ibid., p. 140.

16. See A. C. Crombie, *Robert Grosseteste and the Rise of Experimental Science* (Oxford: Clarendon Press, 1953).

17. Leff, *Paris and Oxford*, p. 199.

18. Ibid., p. 199.

19. Ibid., p. 201.

20. See Gilson, *Reason and Revelation*, pp. 57–59 for an interpretation of this two-truths doctrine.

21. Cited in Gilson, *Reason and Revelation*, p. 63.

22. Leff, *Paris and Oxford*, p. 223.

23. Ibid., pp. 224–225.

24. Privately circulated translation of *Errores Philosophorum* by John Murdoch, in my possession, p. 4.

25. Ibid., p. 4.

26. Leff, *Paris and Oxford*, p. 230.

27. C. Warren Hollister, *Medieval Europe: A Short History*, 4th ed. (New York: Wiley, 1978), p. 273.

28. William of Ockham, *Philosophical Writings*, translated & introduced by Philostheus Bodhner, O.F.M. (Indianapolis: Bobbs-Merrill, 1957), pp. xix, xx, xxii.

29. Ibid., p. xx.

30. Galileo Galilei, *Dialogs Concerning Two New Sciences*, trans. by Henry Crew and Alfonso de Salvio (1638: reprint ed., Evanston, Ill: Northwestern University Press, 1968), pp. 166–167. Emphases mine.

31. See Marshall Clagett, *The Science of Mechanics in the Middle Ages* (Madison: University of Wisconsin Press, 1961), pp. 331–381, for a discussion and partial translation of Oresme. My reduction of Oresme's statements to modern analytic notation introduces an admitted distortion which can be ignored for our purposes, but which may be important for others.

32. Ibid., p. 340.

33. Ockham, *Philosophical Writings*, pp. xlix, 160–163.

CHAPTER 7

1. Cited in Benjamin Farrington, *The Philosophy of Francis Bacon* (Chicago: University of Chicago Press, 1964), p. 27.

2. Cited in Eugenio Garin, *Italian Humanism* (1947; reprint ed., New York: Harper & Row, 1965), p. 25.

3. Ibid., p. 23.

4. *Petrarch's Letters to Classical Authors*, translated by M. E. Cosemya (Chicago: University of Chicago Press, 1910), p. 19.

5. Garin, *Italian Humanism*, p. 20.

6. Francis Bacon, *The Great Instauration*, in *A Selection of His Works*, S. Warhaft, ed. (New York: Odyssey Press, 1965), p. 310.

7. Cited in Benjamin Farrington, *The Philosophy of Francis Bacon* (Chicago: University of Chicago Press, 1964), p. 93.

8. Garin, *Italian Humanism*, p. 23.

9. Ibid., p. 49.

10. Ibid., p. 32.

11. Ibid., p. 35.

12. Ibid., p. 36.

13. Ibid.

14. Ibid., pp. 67–68.

15. Ibid., p. 44.

16. On Pelaconi, Toscanelli, and their relation to medieval optical studies see A. Parronchi, *Studi su Dolce Prospettiva* (Milan: A. Martello 1964), and Madeline Burnside-Lukan, "Alberti to Galileo—The Renaissance Perception of Space" (Ph.D. diss., University of California at Santa Cruz, 1976), pp. 63 ff.

17. See Eugenio Garin, *Science and Civic Life in the Italian Renaissance* (Garden City, N.J.: Anchor Books, 1969), pp. xvii–xix.

18. Giorgio de Santillana, "The Role of Art in the Scientific Renaissance," in Marshall Clagett, *Critical Problems in the History of Science* (Madison: University of Wisconsin Press, 1959), p. 34.

19. Vasari, cited in Paolo Rossi, *Philosophy, Technology, and the Arts in the Early Modern Era* (New York: Harper & Row, 1970), p. 18.

20. De Santillana, "The Role of Art," p. 49.

21. Cited in ibid., p. 35.

22. George Sarton, *Six Wings: Men of Science in the Renaissance* (Bloomington: Indiana University Press, 1957), p. 25.

23. Leon Battesta Alberti, *On Painting* (New Haven: Yale University Press, 1966), p. 90.

24. Ibid.

25. Ibid., p. 46.

26. Ibid., p. 43.

27. Ibid., p. 73.

28. Ibid., pp. 77–78.

29. Ibid., p. 81.

30. See, for example, Marshall Clagett, *Archimedes in the Middle Ages* (Madison: University of Wisconsin Press, 1964), p. 1.

31. A short English language account of the theoretical details of Brunelleschi's solution can be found in William B. Parson's *Engineers and*

Engineering in the Renaissance (1935; reprint ed., Cambridge: MIT Press, 1968), 587–609.

32. Santillana, *The Role of Art*, p. 43.

33. Ibid., p. 44.

34. On Alberti's impact see Bertrand Gille, *Engineers of the Renaissance* (Cambridge: MIT Press, 1966), p. 93.

35. Cited in Rossi, *Philosophy, Technology and Arts*, p. 33.

36. See Stillman Drake and I. E. Drabkin, *Mechanics in the Sixteenth Century* (Madison: University of Wisconsin Press, 1967), for a full discussion of the scientific work of the Italian military engineers.

37. A. R. Hall, *Ballistics in the Seventeenth Century* (Cambridge, England: Cambridge University Press, 1952).

38. Rossi, *Philosophy, Technology and Arts*, pp. 60–61.

39. Ibid., p. 55.

40. Ibid., p. 56.

41. Ibid.

42. In Frederic R. White, *Famous Utopians of the Renaissance* (New York: Hendricks House, 1946), pp. 13, 15.

43. Ibid., p. 48.

44. Drake and Drabkin, *Mechanics*, p. 256.

45. Ibid., p. 256.

46. Cited in Rossi, *Philosophy, Technology and Arts*, pp. 7–8.

47. I have not chosen to focus on navigation and its relationship to mathematics and astronomy. For a good introduction, see Antonia McLean, *Humanism and the Rise of Science in Tudor England* (New York: Neal Watson, 1972), pp. 119 ff.

48. Antonia McLean, *Humanism*, is particularly good on the relation of printing to expanding interest in science during the sixteenth century. See also Elizabeth L. Eisenstein, "Some Conjectures about the Impact of Printing on Western Society and Thought: A Preliminary Report," *Journal of Modern History* (1968), pp. 1–56.

49. Cited in Rossi, *Philosophy, Technology and Arts*, p. 6.

50. Ibid.

CHAPTER 8

1. Elizabeth Eisenstein, "Some Conjectures about the Impact of Printing on Western Society, and Thought: A Preliminary Report," *Journal of Modern History* (1968), pp. 1–56.

2. An English translation by G. R. S. Mead is available as volume 2 of *Thrice-Great Hermes: Studies in Hellenistic Theosophy and Gnosis* (London: John M. Watkins, 1949).

3. Cited in Frances A. Yates, *Giordano Bruno and the Hermetic Tradition* (1964; reprint ed., New York: Vintage Books, 1969), p. 23. All quotations of Hermetic passages are from Yates, whose translations and paraphrases are easier for most students to read than those of the Mead version.

4. Ibid., p. 23.

5. Ibid., pp. 23–24. Emphasis mine.

6. Ibid., pp. 35–36. Emphases mine.

7. Ibid., p. 39.

8. Ibid., p. 41.

9. Cited in Allen Debus, *Man and Nature in the Renaissance* (Cambridge, England: Cambridge University Press, 1978), p. 13.

10. Yates, *Giordano Bruno*, pp. 113–116.

11. Yates presents what is in most respects the best overall discussion of Hermetic magic and its place in Renaissance culture in *Giordano Bruno and the Hermetic Tradition*. A solid, but somewhat simplistic survey of Renaissance astrology and magic can be found in Wayne Schumacher, *The Occult Sciences in the Renaissance* (Berkeley, Los Angeles, London: University of California Press, 1972). A massive amount of information can be found in Lynn Thorndike's *History of Magic and Experimental Science*, vols. IV–VI (New York: Columbia University Press, 1934–41).

12. Cited in Yates, *Giordano Bruno*, pp. 147–148. Emphasis mine.

13. See J. E. L. Dreyer, *Tycho Brahe: A Picture of Scientific Life and Work in the Sixteenth Century* (1890; reprint ed., New York: Dover Publications, 1963), especially pp. 75–79, on Tycho's commitments to astrological doctrines.

14. See Arthur Koestler, *The Watershed: A Biography of Johannes Kepler* (Garden City, N.Y.; Anchor Books, 1960).

15. Debus, *Man and Nature*, p. 21.

16. Cited in Marie Boas, *The Scientific Renaissance* (New York: Harper, 1962), p. 182.

17. Eugenio Garin, *Science and Civic Life in the Italian Renaissance* (Garden City, N.Y.: Anchor Books, 1969), pp. 149–50.

18. Nicole Oresme, *Le livre du Ciel et du monde* (Madison: University of Wisconsin Press, 1968), p. 289.

19. See Robert Westman, "The Astronomers' Role in the Sixteenth Century: A Preliminary Study," *History of Science, 18* (1980), pp. 105–147.

20. Ibid., pp. 118–119.

21. Ibid., pp. 106, 136 n. 6.

22. Cited in Morris Cohen and I. E. Drabkin, *A Sourcebook in Greek Science* (Cambridge: Harvard University Press, 1966), pp. 90–91. Emphases mine.

23. See Edward Rosen, ed., *Three Copernican Treatises* (New York; Dover, 1939), p. 67.

24. See J. R. Ravetz, *Astronomy and Cosmology in the Achievement of Nicolaus Copernicus* (Warsaw; Polish Academy of Sciences, 1965), for a thorough discussion of this problem in the thought of Copernicus.

25. Rosen, *Three Copernican Treatises*, p. 65.

26. From the preface to *On the Revolutions of the Spheres of the Universe*, cited in I. B. Cohen, *The Nature and Growth of the Physical Sciences*, preliminary ed. (New York: Wiley, 1954), p. 87.

27. Rosen, *Three Copernican Treatises*, pp. 58–59.

28. See for example, Westman, "The Astronomers' Role."

29. Rosen, *Three Copernican Treatises*, p. 24. Emphasis mine.

30. Ibid., p. 25.

31. Cited in Marjorie Nicolson, *The Breaking of the Circle: Studies on the Effect of the New Science on Seventeenth Century Poetry* (New York: Columbia University Press, 1960), p. 28.

32. Shakespeare, *The Complete Works*, G. B. Harrison, ed. (New York: Harcourt, Brace, 1948), pp. 983–984. Act I, scene 3.

33. Cited in Nicolson, *Breaking of Circle*, p. 26.

34. See John M. Major, *Sir Thomas Elyot and Renaissance Humanism* (Lincoln: University of Nebraska Press, 1964), p. 17.

35. Thomas Elyot, *The Book Named the Governor*, S. E. Lehmberg, ed. (New York; Dutton, 1962), p. xiii.

36. Cited in E. M. W. Tillyard, *The Elizabethan World Picture* (New York: Vintage), pp. 11–12, 88.

37. Preface to *The General History of the World* (1614) in *The Harvard Classics*, Charles W. Eliot, ed., vol. 39 (New York: PF Collier, 1910), p. 72.

38. Ibid., pp. 97–98.

39. Yates, *Giordano Bruno*, p. 119.

40. Dionysius the Areopagite, *The Mystical Theology and the Celestial Hierarchies* (North Goddming, Surrey: The Shrine of Wisdom, 1949), p. 55.

41. Cited in Dom Denys Rutledge, *Cosmic Theology: The Ecclesiastical Hierarchy of Pseudo-Denys: An Introduction* (London: Routledge and Kegan Paul, 1964), pp. 150–151.

42. Cited in S. F. Mason, "The Scientific Revolution and the Protestant Reformation—I: Calvin and Servitus in Relation to the New Astronomy and the Theory of the Circulation of the Blood," *Annals of Science 9* (1953), 64–81.

43. Ibid.

44. Rosen, *Three Copernican Treatises*, p. 146. Emphasis mine.

45. Cited in Mason, "The Scientific Revolution." Emphasis mine.
46. Ibid. Emphasis mine.
47. Richard Bostock, cited in Mason, "The Scientific Revolution."
48. Cited in Thomas Kuhn, *The Copernican Revolution* (New York: Vintage, 1957), p. 192.
49. Cited in S. F. Mason, "The Scientific Revolution."
50. Stillman Drake, *Galileo at Work: His Scientific Biography* (Chicago: University of Chicago Press, 1978), p. 225. Emphasis mine.
51. For a fascinating and beautifully written account of Galileo's trial, see Giorgia de Santillana, *The Crime of Galileo* (Chicago: University of Chicago Press, 1955).
52. Ibid., p. 312.
53. Ibid., p. 305.
54. John Milton, *Areopagetica* in *Complete Poems and Major Prose*, Merritt Y. Hughes, ed. (Indianapolis: The Odyssey Press, 1957), pp. 737–738.
55. On the centrality of Jesuit scholars to seventeenth-century science, see especially John Heilbron, *Electricity in the 17th and 18th Centuries* (Berkeley, Los Angeles, London: University of California Press, 1979), pp. 101–114.

CHAPTER 9

1. The extent of Andreae's mysticism and his role in the Rosicrucian movement of the early seventeenth century are topics of intense scholarly debate. In what follows I accept the interpretations of John Warwick Montgomery, whose two-volume *Cross and Crucible: John Valentine Andreae (1586–1654)* (The Hague: Martinus Nijhoff, 1973), is the most extensive English language treatment of Andreae's life and ideas.
2. See Montgomery, *Cross and Crucible*, Vol. II for a full analysis of *The Chemical Wedding*.
3. On the Rosicrucian movement and its links to seventeenth century scientific development, see Frances Yates, *The Rosicrucian Enlightenment* (London: Routledge and Kegan Paul, 1972).
4. Montgomery, *Cross and Crucible*, 1:46.
5. Ibid., p. 46.
6. Ibid.
7. Cited in ibid., pp. 135–36.
8. Ibid., p. 51. Latinists may disagree with my translation of the final line, which reads, "Deo vero Omnipotenti, uni, soli, semper gloria esto." "Soli" here could either mean the sun or alone. My choice is based partly on my wish to see God identified with the sun and partly on the grounds that the usual formulaic way to describe God as "the one and only" is to use the form "uni et soli," rather than to isolate the two terms

as Andreae intentionally seems to do. It may well be that Andreae's passage involves an intentional play on words.

9. See Montgomery, *Cross and Crucible*, 1:68–69.

10. Cited in ibid., pp. 218–219.

11. Felix E. Held, *Christianopolis: An Ideal State of the Seventeenth Century* (Oxford: Oxford University Press, 1916), pp. 41–74, 100–126.

12. Though Held published an English translation of Christianopolis in 1916, the book is rare. Interested students can find important passages summarized in Lewis Mumford, *The Story of Utopia* (New York: Viking Press, 1922), pp. 81–99.

13. Held, *Christianopolis*, p. 142.

14. Ibid., p. 145.

15. Ibid., p. 148.

16. Ibid., pp. 154–155. Emphasis mine.

17. Ibid., p. 196.

18. Ibid., p. 199.

19. Ibid., p. 202.

20. Ibid., pp. 220–221.

21. Ibid., p. 228.

22. Ibid., p. 229.

23. Ibid., pp. 231–232. Emphasis mine.

24. Cited in Davis D. McElroy, *Scotland's Age of Improvement* (Seattle: Washington University Press, 1969), p. 6.

25. Cited in Benjamin Farrington, *Francis Bacon: Philosopher of Industrial Science* (1949; reprint ed., New York: Collier Books, 1961), p. 121.

26. Ibid., p. 118.

27. Cited in Benjamin Farrington, *The Philosophy of Francis Bacon* (Chicago: University of Chicago Press, 1964), p. 29.

28. Francis Bacon, *Novum Organum*, Book I, Aphorism 67, in *A Selection of His Writings* (New York: Odyssey Press, 1965), p. 348.

29. Ibid., Aphorism 129, p. 374.

30. Ibid.

31. Ibid.

32. Francis Bacon, *New Atlantis*, in *A Selection of His Writings*, p. 447.

33. Cited in Benjamin Farrington, *The Philosophy of Francis Bacon*, pp. 39, 19.

34. Bacon, *The Proficience and Advancement of Learning* (1605), in *A Selection of His Writings*, p. 229.

35. Ibid.

36. Bacon, *Novum Organum*, Book I, Aphorisms 23–69, contains Bacon's

best and most extensive discussion of why all other philosophers had been led into error.

37. On Palissy and Bacon see Paolo Rossi, *Francis Bacon: From Magic to Science* (London: Routledge and Kegan Paul, 1968), pp. 8–9.

38. Bacon, Proemium to *The Great Instauration* (1620), in *A Selection of His Writings*, p. 299.

39. Ibid., p. 314.

40. Farrington, *The Philosophy of Francis Bacon*, p. 32.

41. Bacon, *New Atlantis*, in *A Selection of His Writings*, p. 421.

42. Ibid.

43. Ibid., p. 427. Emphasis Bacon's.

44. Ibid., p. 418.

45. Ibid., p. 447.

46. Ibid., p. 450.

47. Ibid., pp. 451–454.

48. Ibid., p. 456.

49. See Peter Mathias, "Who Unbound Prometheus? Science and Technical Change, 1600–1800," in Peter Mathias, ed., *Science and Society: 1600–1900* (Cambridge, England: Cambridge University Press, 1972), especially pp. 74–76.

50. See Richard Foster Jones, *The Ancients and the Moderns*, 2d ed. (Seattle: University of Washington Press, 1961), pp. 87–118.

51. See Roger Hahn, *The Anatomy of a Scientific Institution: The Paris Academy of Sciences, 1666–1803* (Berkeley, Los Angeles, London: University of California Press, 1970), pp. 8–18.

52. Samuel Johnson, *Idler, 88* (December 22, 1759).

53. Emile de Bois-Reymond, *Kulturegeschichte und Naturewissenschaften* (1878) cited in Roland N. Stromberg, ed., *Realism, Naturalism, and Symbolism* (New York: Harper & Row, 1968), p. 29.

54. Ibid., p. 30.

Index

Designer: Linda Robertson
Compositor: Trend Western
Printer: Vail-Ballou
Binder: Vail-Ballou
Text: Auriga
Display: Auriga and Helvetica